SpringerBriefs in Mathematics

For other titles published in this series, go to
http://www.springer.com/series/10030

George A. Anastassiou

Inequalities Based on Sobolev Representations

 Springer

George A. Anastassiou
Department of Mathematical Sciences
University of Memphis
Memphis, TN 38152
USA
ganastss@memphis.edu

ISSN 2191-8198   e-ISSN 2191-8201
ISBN 978-1-4614-0200-8   e-ISBN 978-1-4614-0201-5
DOI 10.1007/978-1-4614-0201-5
Springer New York Dordrecht Heidelberg London

Library of Congress Control Number: 2011929870

Mathematics Subject Classification (2010): 26D10, 26D15, 26D20

Printed on acid-free paper

Springer is part of Springer Science+Business Media (www.springer.com)

To the memory of my friend Khalid Khouri who left this world too young

The measure of success for a person is the magnitude of his/her ability to convert negative conditions to positive ones and achieve goals
    —The author

Preface

This brief monograph is the first one to deal exclusively with very general tight integral inequalities of Chebyshev–Grüss, Ostrowski types, and of the comparison of integral means. These rely on the well-known Sobolev integral representations of functions. The inequalities engage ordinary and weak partial derivatives of the involved functions. Applications of these developments are illustrated. On the way to prove the main results we derive important estimates for the averaged Taylor polynomials and remainders of Sobolev integral representations. The exposed results expand to all possible directions. We examine both the univariate and multivariate cases.

For the convenience of the reader, each chapter of this book is written in a self-contained style.

This treatise relies on the author's last year of related research work.

Advanced courses and seminars can be taught out of this brief book. All necessary background and motivations are given in each chapter. A related list of references is also given at the end of each chapter. These results first appeared in my articles that are mentioned in the references. The results are expected to find applications in many subareas of mathematical analysis, inequalities, partial differential equations, information theory, etc. As such this monograph is suitable for researchers, graduate students, seminars of the above subjects, and also for all science libraries.

The preparation of this booklet took place during 2010–2011 in Memphis, TN, USA.

I thank my family for their dedication and love to me, which was the strongest support during the writing of this book.

Department of Mathematical Sciences   George A. Anastassiou
The University of Memphis, TN, USA
March 5, 2011

Contents

Chapter 1
Univariate Integral Inequalities Based on Sobolev Representations

Here we present very general univariate tight integral inequalities of Cheby-shev–Grüss, Ostrowski types for comparison of integral means and information theory. These are based on the well-known Sobolev integral representations of a function. The inequalities engage ordinary and weak derivatives of the involved functions. Applications are given. On the way to prove the main results we derive important estimates for the averaged Taylor polynomials and remainders of Sobolev integral representations. The results are explained thoroughly. This chapter relies on [4].

1.1 Introduction

This chapter is greatly motivated by the following theorems:

Theorem 1.1 (Chebychev, 1882, [7]). *Let* $f, g: [a, b] \to \mathbb{R}$ *absolutely continuous functions. If* $f', g' \in L_\infty ([a, b])$, *then*

$$\left| \frac{1}{b-a} \int_a^b f(x) g(x) \, dx - \frac{1}{(b-a)^2} \left(\int_a^b f(x) \, dx \right) \left(\int_a^b g(x) \, dx \right) \right|$$

$$\leq \frac{1}{12} (b-a)^2 \|f'\|_\infty \|g'\|_\infty . \tag{1.1}$$

Theorem 1.2 (G. Grüss, 1935, [11]). *Let* f, g *integrable functions from* $[a, b] \to \mathbb{R}$, *such that* $m \leq f(x) \leq M$, $\rho \leq g(x) \leq \sigma$, *for all* $x \in [a, b]$, *where* $m, M, \rho, \sigma \in \mathbb{R}$. *Then*

$$\left| \frac{1}{b-a} \int_a^b f(x) g(x) \, dx - \frac{1}{(b-a)^2} \left(\int_a^b f(x) \, dx \right) \left(\int_a^b g(x) \, dx \right) \right|$$

$$\leq \frac{1}{4} (M - m) (\sigma - \rho) . \tag{1.2}$$

In 1938, A. Ostrowski [14] proved.

G.A. Anastassiou, *Inequalities Based on Sobolev Representations*,
SpringerBriefs in Mathematics 2, DOI 10.1007/978-1-4614-0201-5_1,
© George A. Anastassiou

Theorem 1.3. *Let $f : [a,b] \to \mathbb{R}$ be continuous on $[a,b]$ and differentiable on (a,b) whose derivative $f' : (a,b) \to \mathbb{R}$ is bounded on (a,b), i.e., $\|f'\|_\infty = \sup_{t \in (a,b)} |f'(t)| < +\infty$. Then*

$$\left| \frac{1}{b-a} \int_a^b f(t)\,dt - f(x) \right| \leq \left[\frac{1}{4} + \frac{\left(x - \frac{a+b}{2}\right)^2}{(b-a)^2} \right] \cdot (b-a) \, \|f'\|_\infty, \quad (1.3)$$

for any $x \in [a,b]$. The constant $\frac{1}{4}$ is the best possible.

See also [1–3] for related works that inspired as well this chapter.

In this chapter using the univariate Sobolev-type representation formulae, see Theorems 1.10, 1.14 and also Corollaries 1.11, 1.12, we first estimate their remainders and then the involved averaged Taylor polynomials.

Based on these estimates we derive lots of very tight inequalities on $\mathbb{R}$: of Chebyshev–Grüss type, Ostrowski type, for comparison of integral means and Csiszar's f-divergence with applications. The results involve ordinary and weak derivatives and they go to all possible directions using various norms. All of our tools come from the excellent monograph by V. Burenkov, [6].

1.2 Background

Here we follow [6].

For a measurable nonempty set $\Omega \subset \mathbb{R}^n$, $n \in \mathbb{N}$ we shall denote by $L_p^{\text{loc}}(\Omega)$ $(1 \leq p \leq \infty)$ – the set of functions defined on Ω such that for each compact $K \subset \Omega$, $f \in L_p(K)$.

Definition 1.4. Let $\Omega \subset \mathbb{R}^n$ be an open set, $\alpha \in \mathbb{Z}_+^n$, $\alpha \neq 0$ and $f, g \in L_1^{\text{loc}}(\Omega)$. The function g is a weak derivative of the function f of order α on Ω (briefly $g = D_w^\alpha f$) if $\forall \, \varphi \in C_0^\infty(\Omega)$ (i.e., $\varphi \in C^\infty(\Omega)$ compactly supported in Ω)

$$\int_\Omega f D^\alpha \varphi \, dx = (-1)^{|\alpha|} \int_\Omega g\varphi \, dx. \quad (1.4)$$

Definition 1.5. $W_p^l(\Omega)$ $(l \in \mathbb{N}, 1 \leq p \leq \infty)$ – Sobolev space, which is the Banach space of functions $f \in L_p(\Omega)$ such that $\forall \, \alpha \in \mathbb{Z}_+^n$ where $|\alpha| \leq l$ the weak derivatives $D_w^\alpha f$ exist on Ω and $D_w^\alpha f \in L_p(\Omega)$, with the norm

$$\|f\|_{W_p^l(\Omega)} = \sum_{|\alpha| \leq l} \|D_w^\alpha f\|_{L_p(\Omega)}. \quad (1.5)$$

Definition 1.6. For $l \in \mathbb{N}$, we define the Sobolev-type local space $\left(W_1^l\right)^{(\text{loc})}(\Omega) := \{f : \Omega \to \mathbb{R} : f \in L_{\text{loc}}^1(\Omega)$ and all f-distributional partials of orders $\leq l$ belong to $L_{\text{loc}}^1(\Omega)\} = \{f \in L_1^{\text{loc}}(\Omega) :$ for each open set G compactly embedded into Ω, $f \in W_1^l(G)\}$.

We use Definitions 1.4–1.6 on $\mathbb{R}$. Next we mention Sobolev's integral representation from [6].

Definition 1.7 ([6], p. 82). Let $-\infty < a < b < \infty$,

$$\omega \in L_1(a,b), \quad \int_a^b \omega(x)\,dx = 1. \tag{1.6}$$

Define

$$\Lambda(x,y) := \begin{cases} \int_a^y \omega(u)\,du, & a \le y \le x \le b, \\ -\int_y^b \omega(u)\,du, & a \le x < y \le b. \end{cases} \tag{1.7}$$

Proposition 1.8 ([6], p. 82). *Let f be absolutely continuous on $[a,b]$. Then $\forall\, x \in (a,b)$*

$$f(x) = \int_a^b f(y)\,\omega(y)\,dy + \int_a^b \Lambda(x,y)\,f'(y)\,dy, \tag{1.8}$$

the simplest case of Sobolev's integral representation.

Remark 1.9 ([6], pp. 82–83). We have that Λ is bounded:

$$\forall\, x, y \in [a,b], \quad |\Lambda(x,y)| \le \|\omega\|_{L_1(a,b)}, \tag{1.9}$$

and if $\omega \ge 0$, then

$$\forall\, x, y \in [a,b], \quad |\Lambda(x,y)| \le \Lambda(b,b) = 1.$$

If ω is symmetric with respect to $\frac{a+b}{2}$, then $\forall\, y \in [a,b]$ we have

$$\left| \Lambda\left(\frac{a+b}{2}, y \right) \right| \le \frac{1}{2}.$$

Examples of ω:

$$\omega(x) = \frac{1}{b-a}, \quad \forall x \in (a,b),$$

also

$$\omega(x) = \frac{1}{2m}\left(\chi_{(a,a+\frac{1}{m})} + \chi_{(b-\frac{1}{m},b)} \right),$$

where $\chi_{(\alpha,\beta)}$ denotes the characteristic function of (α,β), $m \in \mathbb{N}$ and $m \ge 2(b-a)^{-1}$.

If $f \in (W_1^l)^{\text{loc}}(a,b)$, then f is equivalent to a function, which is locally absolutely continuous on (a,b) (its ordinary derivative, which exists almost everywhere on (a,b), is a weak derivative f_w' of f). Thus (1.8) holds for almost every $x \in (a,b)$ if f' is replaced by f_w'.

In this chapter sums of the form $\sum_{k=1}^{0} \cdot = 0$.
We mention

Theorem 1.10 ([6], p. 83). *Let* $l \in \mathbb{N}$, $-\infty \le a < \alpha < \beta < b \le \infty$ *and*

$$\begin{cases} \omega \in L_1(\mathbb{R}), & (support) \sup p\omega \subset [\alpha, \beta], \\ \int_{\mathbb{R}} \omega(x)\,dx = 1. \end{cases} \tag{1.10}$$

Moreover, assume that the derivative $f^{(l-1)}$ *exists and is locally absolutely contin-uous on* (a,b). *Then* $\forall\, x \in (a,b)$

$$f(x) = \sum_{k=0}^{l-1} \frac{1}{k!} \int_a^b f^{(k)}(y)(x-y)^k \omega(y)\,dy$$

$$+ \frac{1}{(l-1)!} \int_a^b (x-y)^{l-1} \Lambda(x,y) f^{(l)}(y)\,dy, \tag{1.11}$$

and

$$f(x) = \sum_{k=0}^{l-1} \frac{1}{k!} \int_\alpha^\beta f^{(k)}(y)(x-y)^k \omega(y)\,dy$$

$$+ \frac{1}{(l-1)!} \int_{a_x}^{b_x} (x-y)^{l-1} \Lambda(x,y) f^{(l)}(y)\,dy, \tag{1.12}$$

where $a_x = x$, $b_x = \beta$ *for* $x \in (a,\alpha]$; $a_x = \alpha$, $b_x = \beta$ *for* $x \in (\alpha,\beta)$; $a_x = \alpha$, $b_x = x$ *for* $x \in [\beta,b)$.

If, in particular, $-\infty < a < b < \infty$, $f^{(l-1)}$ *exists and is absolutely continuous on* $[a,b]$, *then (1.11), (1.12) hold* $\forall\, x \in [a,b]$ *and for any interval* $(\alpha,\beta) \subset (a,b)$.

Corollary 1.11 ([6], p. 85). *Assume that* $l > 1$, *condition (1.10) is replaced by*

$$\begin{cases} \omega \in C^{(l-2)}(\mathbb{R}), & \sup p\omega \subset [\alpha,\beta], \\ \int_{\mathbb{R}} \omega(x)\,dx = 1, \end{cases} \tag{1.13}$$

and the derivative $\omega^{(l-2)}$ *is absolutely continuous on* $[a,b]$.
Then for the same f *as in Theorem 1.10,* $\forall\, x \in (a,b)$

$$f(x) = \int_\alpha^\beta \left(\sum_{k=0}^{l-1} \frac{(-1)^k}{k!} \left[(x-y)^k \omega(y) \right]_y^{(k)} \right) f(y)\,dy$$

$$+ \frac{1}{(l-1)!} \int_{a_x}^{b_x} (x-y)^{l-1} \Lambda(x,y) f^{(l)}(y)\,dy. \tag{1.14}$$

In particular, here

$$\omega(\alpha) = \dots = \omega^{(l-2)}(\alpha) = \omega(\beta) = \dots = \omega^{(l-2)}(\beta) = 0. \quad (1.15)$$

Corollary 1.12 ([6], p. 86). *Assume that* $l, m \in \mathbb{N},\ m < l.$ *Then for the same* f *and* ω *as in Corollary 1.11,* $\forall\ x \in (a, b)$

$$f^{(m)}(x) = \int_\alpha^\beta \left(\sum_{k=0}^{l-m-1} \frac{(-1)^{k+m}}{k!} \left[(x-y)^k\, \omega(y) \right]_y^{(k+m)} \right) f(y)\, dy$$

$$+ \frac{1}{(l-m-1)!} \int_{a_x}^{b_x} (x-y)^{l-m-1}\, \Lambda(x, y)\, f^{(l)}(y)\, dy. \quad (1.16)$$

Remark 1.13 ([6], p. 86). The first summand in (1.14) can take the form:

$$\begin{cases} \int_\alpha^\beta \left(\sum_{s=0}^{l-1} \sigma_s\, (x-y)^s\, \omega^{(s)}(y) \right) f(y)\, dy, \\[2mm] \text{where } \sigma_s := \frac{(-1)^s}{s!} \sum_{k=s}^{l-s-1} \binom{s+k}{k}. \end{cases} \quad (1.17)$$

Similarly, we have for the first summand of (1.16) the following form:

$$\begin{cases} \int_\alpha^\beta \left(\sum_{s=m}^{l-1} \sigma_{s,m}\, (x-y)^{s-m}\, \omega^{(s)}(y) \right) f(y)\, dy, \\[2mm] \text{where } \sigma_{s,m} := \frac{(-1)^s}{(s-m)!} \sum_{k=s}^{l-s-1} \binom{s+k}{k}. \end{cases} \quad (1.18)$$

We need

Theorem 1.14 ([6], p. 91). *Let* $l \in \mathbb{N},\ -\infty \le a < \alpha < \beta < b \le \infty,\ \omega$ *satisfy condition*

$$\omega \in L_1(\mathbb{R}),\ \sup p\omega \subset [\alpha, \beta],\ \int_\mathbb{R} \omega(x)\, dx = 1 \quad (1.19)$$

and $f \in \left(W_1^l \right)^{loc}(a, b).$ *Then for almost every* $x \in (a, b)$

$$f(x) = \sum_{k=0}^{l-1} \frac{1}{k!} \int_\alpha^\beta f_w^{(k)}(y)\, (x-y)^k\, \omega(y)\, dy$$

$$+ \frac{1}{(l-1)!} \int_{a_x}^{b_x} (x-y)^{l-1}\, \Lambda(x, y)\, f_w^{(l)}(y)\, dy, \quad (1.20)$$

where a_x, b_x *as in Theorem 1.10.*
   We denote $f_w^{(0)} := f.$

Remark 1.15 ([6], p. 92). By Theorem 1.14 it follows that if in Corollaries 1.11, 1.12 $f \in \left(W_1^l\right)^{loc}(a, b)$ then equalities (1.14) and (1.16) hold almost everywhere on (a, b), if we substitute $f^{(l)}, f^{(m)}$ by the weak derivatives $f_w^{(l)}, f_w^{(m)}$; respectively.

Next we estimate the remainders of the above mentioned Sobolev representations. We make

Remark 1.16. Denote by $\overline{f}^{(k)}$ either $f^{(k)}$ or $f_w^{(k)}$, where $k \in \mathbb{N}$. Let $0 \le m < l$, $m \in \mathbb{Z}_+$. We estimate

$$R_{m,l} f(x) := \frac{1}{(l-m-1)!} \int_\alpha^\beta (x-y)^{l-m-1} \Lambda(x, y) \overline{f}^{(l)}(y) \, dy, \qquad (1.21)$$

for $x \in (\alpha, \beta)$, where Λ as in (1.7), see also (1.9).
    So we obtain

$$R_{0,l} f(x) := \frac{1}{(l-1)!} \int_\alpha^\beta (x-y)^{l-1} \Lambda(x, y) \overline{f}^{(l)}(y) \, dy. \qquad (1.22)$$

Thus we derive

$$|R_{m,l} f(x)| \le \frac{\|\omega\|_{L_1(a,b)} \cdot (\beta - \alpha)^{l-m-1}}{(l-m-1)!} \int_\alpha^\beta \left|\overline{f}^{(l)}(y)\right| dy$$

$$= \frac{\|\omega\|_{L_1(a,b)} \cdot \left\|\overline{f}^{(l)}\right\|_{L_1(\alpha,\beta)} \cdot (\beta - \alpha)^{l-m-1}}{(l-m-1)!}, \qquad (1.23)$$

$x \in (\alpha, \beta)$.
    We also have

$$|R_{m,l} f(x)| \le \frac{\|\omega\|_{L_1(a,b)}}{(l-m-1)!} \int_\alpha^\beta |x-y|^{l-m-1} \left|\overline{f}^{(l)}(y)\right| dy =: I_1.$$

If $\overline{f}^{(l)} \in L_\infty(\alpha, \beta)$, then

$$I_1 \le \frac{\|\omega\|_{L_1(a,b)} \left\|\overline{f}^{(l)}\right\|_{L_\infty(\alpha,\beta)}}{(l-m-1)!} \left(\int_\alpha^\beta |x-y|^{l-m-1} \, dy \right).$$

But

$$\int_\alpha^\beta |x-y|^{l-m-1} \, dy = \int_\alpha^x (x-y)^{l-m-1} \, dy + \int_x^\beta (y-x)^{l-m-1} \, dy$$

$$= \frac{(\beta-x)^{l-m} + (x-\alpha)^{l-m}}{l-m}.$$

Therefore if $\overline{f}^{(l)} \in L_\infty (\alpha, \beta)$, then

$$|R_{m,l} f (x)| \le \frac{\|\omega\|_{L_1(a,b)} \left\|\overline{f}^{(l)}\right\|_{L_\infty (\alpha,\beta)}}{(l - m)!} \left((\beta - x)^{l-m} + (x - \alpha)^{l-m}\right), \quad (1.24)$$

$x \in (\alpha, \beta)$.

Let now $p, q > 1 : \frac{1}{p} + \frac{1}{q} = 1$. If $\overline{f}^{(l)} \in L_p (\alpha, \beta)$, then

$$I_1 \le \frac{\|\omega\|_{L_1(a,b)}}{(l - m - 1)!} \left(\int_\alpha^\beta |x - y|^{q(l-m-1)} \, dy\right)^{\frac{1}{q}} \left\|\overline{f}^{(l)}\right\|_{L_p(\alpha,\beta)}.$$

However

$$\int_\alpha^\beta |x - y|^{q(l-m-1)} \, dy = \int_\alpha^x (x - y)^{q(l-m-1)} \, dy + \int_x^\beta (y - x)^{q(l-m-1)} \, dy$$

$$= \frac{(x - \alpha)^{q(l-m-1)+1} + (\beta - x)^{q(l-m-1)+1}}{q (l - m - 1) + 1}.$$

Hence if $\overline{f}^{(l)} \in L_p (\alpha, \beta)$, then

$$|R_{m,l} f (x)| \le \frac{\|\omega\|_{L_1(a,b)} \left\|\overline{f}^{(l)}\right\|_{L_p(\alpha,\beta)}}{(l - m - 1)!}$$

$$\times \left(\frac{(\beta - x)^{q(l-m-1)+1} + (x - \alpha)^{q(l-m-1)+1}}{q (l - m - 1) + 1}\right)^{\frac{1}{q}}, \quad (1.25)$$

$x \in (\alpha, \beta)$.

If $\sup p\omega \subset [\alpha, \beta]$, then

$$\|\omega\|_{L_1(a,b)} = \|\omega\|_{L_1(\alpha,\beta)}. \quad (1.26)$$

If $\omega \in C (\mathbb{R})$ and $\sup p\omega \subset [\alpha, \beta]$, then

$$\|\omega\|_{L_1(\alpha,\beta)} \le \|\omega\|_{\infty,[\alpha,\beta]} \cdot (\beta - \alpha). \quad (1.27)$$

We make

Remark 1.17. Here we estimate from the Taylor's averaged polynomial, see (1.12) and (1.20), that part

$$Q^{l-1} f(x) := \sum_{k=1}^{l-1} \frac{1}{k!} \int_\alpha^\beta \overline{f}^{(k)}(y)(x-y)^k \, \omega(y) \, dy, \qquad (1.28)$$

called also quasi-averaged Taylor polynomial. When $l = 1$, then $Q^0 f(x) = 0$.
  We see that

$$\left| Q^{l-1} f(x) \right| \le \sum_{k=1}^{l-1} \frac{1}{k!} \int_\alpha^\beta \left| \overline{f}^{(k)}(y) \right| |x-y|^k \, |\omega(y)| \, dy$$

$$\le \left(\sum_{k=1}^{l-1} \frac{(\beta-\alpha)^k}{k!} \left\| \overline{f}^{(k)} \right\|_{L_1(\alpha,\beta)} \right) \|\omega\|_{L_\infty(\mathbb{R})}, \qquad (1.29)$$

given that $\|\omega\|_{L_\infty(\mathbb{R})} < \infty$, $x \in (\alpha, \beta)$.
  Similarly, when $\overline{f}^{(k)} \in L_\infty(\alpha, \beta)$, $k = 1, ..., l-1$, and $\|\omega\|_{L_\infty(\mathbb{R})} < \infty$ we obtain

$$\left| Q^{l-1} f(x) \right| \le \sum_{k=1}^{l-1} \frac{\left\| \overline{f}^{(k)} \right\|_{L_\infty(\alpha,\beta)} \|\omega\|_{L_\infty(\mathbb{R})}}{k!} \int_\alpha^\beta |x-y|^k \, dy$$

$$= \left(\sum_{k=1}^{l-1} \left(\frac{(\beta-x)^{k+1} + (x-\alpha)^{k+1}}{(k+1)!} \right) \left\| \overline{f}^{(k)} \right\|_{L_\infty(\alpha,\beta)} \right) \|\omega\|_{L_\infty(\mathbb{R})}, \qquad (1.30)$$

$x \in (\alpha, \beta)$.
  Let $p, q > 1 : \frac{1}{p} + \frac{1}{q} = 1$, $\overline{f}^{(k)} \in L_p(\alpha, \beta)$, $k = 1, ..., l-1$, and again $\|\omega\|_{L_\infty(\mathbb{R})} < \infty$. Then

$$\left| Q^{l-1} f(x) \right| \le \sum_{k=1}^{l-1} \frac{\left\| \overline{f}^{(k)} \right\|_{L_p(\alpha,\beta)}}{k!} \left(\int_\alpha^\beta |x-y|^{kq} \, dy \right)^{\frac{1}{q}} \|\omega\|_{L_\infty(\mathbb{R})}$$

$$= \left(\sum_{k=1}^{l-1} \left(\frac{(\beta-x)^{(kq+1)} + (x-\alpha)^{(kq+1)}}{kq+1} \right)^{\frac{1}{q}} \frac{\left\| \overline{f}^{(k)} \right\|_{L_p(\alpha,\beta)}}{k!} \right) \|\omega\|_{L_\infty(\mathbb{R})}, \qquad (1.31)$$

$x \in (\alpha, \beta)$.

Suppose $\omega \in L_1(\mathbb{R})$ and $\overline{f}^{(k)} \in L_\infty(\alpha, \beta)$, $k = 1, ..., l-1$, then

$$\left| Q^{l-1} f(x) \right| \leq \left(\sum_{k=1}^{l-1} \frac{\left\| \overline{f}^{(k)} \right\|_{L_\infty(\alpha,\beta)} (\beta - \alpha)^k}{k!} \right) \| \omega \|_{L_1(\mathbb{R})}, \qquad (1.32)$$

$x \in (\alpha, \beta)$.

Suppose $p, q > 1 : \frac{1}{p} + \frac{1}{q} = 1$, $\overline{f}^{(k)} \in L_p(\alpha, \beta)$, $k = 1, ..., l-1$; $\omega \in L_q(\alpha, \beta)$, then

$$\left| Q^{l-1} f(x) \right| \leq \left(\sum_{k=1}^{l-1} \frac{(\beta - \alpha)^k}{k!} \left\| \overline{f}^{(k)} \right\|_{L_p(\alpha,\beta)} \right) \| \omega \|_{L_q(\alpha,\beta)}, \qquad (1.33)$$

$x \in (\alpha, \beta)$.

Suppose $p, q, r > 1 : \frac{1}{p} + \frac{1}{q} + \frac{1}{r} = 1$, $\overline{f}^{(k)} \in L_p(\alpha, \beta)$, $k = 1, ..., l-1$; $\omega \in L_q(\alpha, \beta)$, then

$$\left| Q^{l-1} f(x) \right| \leq \left(\sum_{k=1}^{l-1} \frac{\left\| \overline{f}^{(k)} \right\|_{L_p(\alpha,\beta)}}{k!} \left(\frac{(\beta - x)^{(kr+1)} + (x - \alpha)^{(kr+1)}}{(kr + 1)} \right)^{\frac{1}{r}} \right)$$

$$\times \| \omega \|_{L_q(\alpha,\beta)}, \qquad (1.34)$$

$x \in (\alpha, \beta)$.

We also make

Remark 1.18. Here $l > 1$, $\omega \in C^{(l-2)}(\mathbb{R})$, sup $p\omega \subset [\alpha, \beta]$, $\int_\mathbb{R} \omega(x) \, dx = 1$, and the derivative $\omega^{(l-2)}$ is absolutely continuous on $[a, b]$. Hence we have that

$$Q^{l-1} f(x) = \sum_{k=1}^{l-1} \frac{(-1)^k}{k!} \int_\alpha^\beta \left(\left[(x - y)^k \omega(y) \right]_y^{(k)} \right) f(y) \, dy, \qquad (1.35)$$

$\forall \, x \in (\alpha, \beta)$.

And it holds

$$\left| Q^{l-1} f(x) \right| \leq \sum_{k=1}^{l-1} \frac{1}{k!} \int_\alpha^\beta \left| \left[(x - y)^k \omega(y) \right]_y^{(k)} \right| |f(y)| \, dy, \qquad (1.36)$$

$\forall \, x \in (\alpha, \beta)$.

Consequently, $\forall \, x \in (\alpha, \beta)$,

$$
\left| Q^{l-1} f \left(x \right) \right|
$$

$$
\leq
\begin{cases}
\left(\sum_{k=1}^{l-1} \frac{1}{k!} \left\| \left[(x-y)^k \, \omega \left(y \right) \right]_y^{(k)} \right\|_\infty \right) \| f \|_{L_1(\alpha,\beta)}, \\
\quad \text{if } f \in L_1 \left(\alpha, \beta \right), \\
\left(\sum_{k=1}^{l-1} \frac{1}{k!} \left\| \left[(x-y)^k \, \omega \left(y \right) \right]_y^{(k)} \right\|_{L_1(\alpha,\beta)} \right) \| f \|_{L_\infty(\alpha,\beta)}, \\
\quad \text{if } f \in L_\infty \left(\alpha, \beta \right), \\
\quad \text{when } p, q > 1 : \frac{1}{p} + \frac{1}{q} = 1, \text{ we have} \\
\left(\sum_{k=1}^{l-1} \frac{1}{k!} \left\| \left[(x-y)^k \, \omega \left(y \right) \right]_y^{(k)} \right\|_{L_q(\alpha,\beta)} \right) \| f \|_{L_p(\alpha,\beta)}, \\
\quad \text{if } f \in L_p \left(\alpha, \beta \right).
\end{cases}
\tag{1.37}
$$

Let $l, m \in \mathbb{N}$, $m < l$, and f, ω as above, $x \in (\alpha, \beta)$.
We consider here

$$
Q_m^{l-1} f \left(x \right) := \sum_{k=1}^{l-m-1} \frac{(-1)^{k+m}}{k!} \int_\alpha^\beta \left(\left[(x-y)^k \, \omega \left(y \right) \right]_y^{(k+m)} \right) f \left(y \right) \mathrm{d}y. \tag{1.38}
$$

When $l = m + 1$, then $Q_m^{l-1} f \left(x \right) := 0$.
Hence it holds

$$
\left| Q_m^{l-1} f \left(x \right) \right| \leq \sum_{k=1}^{l-m-1} \frac{1}{k!} \int_\alpha^\beta \left| \left(\left[(x-y)^k \, \omega \left(y \right) \right]_y^{(k+m)} \right) \right| \left| f \left(y \right) \right| \mathrm{d}y, \tag{1.39}
$$

$\forall \, x \in (\alpha, \beta)$.
Consequently, $\forall \, x \in (\alpha, \beta)$,

$$
\left| Q_m^{l-1} f \left(x \right) \right|
$$

$$
\leq
\begin{cases}
\left(\sum_{k=1}^{l-m-1} \frac{1}{k!} \left\| \left[(x-y)^k \, \omega \left(y \right) \right]_y^{(k+m)} \right\|_\infty \right) \| f \|_{L_1(\alpha,\beta)}, \\
\quad \text{if } f \in L_1 \left(\alpha, \beta \right), \\
\left(\sum_{k=1}^{l-m-1} \frac{1}{k!} \left\| \left[(x-y)^k \, \omega \left(y \right) \right]_y^{(k+m)} \right\|_{L_1(\alpha,\beta)} \right) \| f \|_{L_\infty(\alpha,\beta)}, \\
\quad \text{if } f \in L_\infty \left(\alpha, \beta \right), \\
\quad \text{when } p, q > 1 : \frac{1}{p} + \frac{1}{q} = 1, \text{ we have} \\
\left(\sum_{k=1}^{l-m-1} \frac{1}{k!} \left\| \left[(x-y)^k \, \omega \left(y \right) \right]_y^{(k+m)} \right\|_{L_q(\alpha,\beta)} \right) \| f \|_{L_p(\alpha,\beta)}, \\
\quad \text{if } f \in L_p \left(\alpha, \beta \right).
\end{cases}
\tag{1.40}
$$

We also need

Remark 1.19. Here again $\overline{f}^{(k)}$ means either $f^{(k)}$ or $f_w^{(k)}$, $k \in \mathbb{N}$. We rewrite (1.12), (1.14), and (1.20). For $x \in (\alpha, \beta)$ we get

$$f(x) = \int_\alpha^\beta f(y) \omega(y) \, dy + Q^{l-1} f(x) + R_{0,l} f(x). \qquad (1.41)$$

Also for $x \in (\alpha, \beta)$ we rewrite (1.16) (see also Remark 1.15) as follows:

$$\overline{f}^{(m)}(x) = (-1)^m \int_\alpha^\beta f(y) \omega^{(m)}(y) \, dy + Q_m^{l-1} f(x) + R_{m,l} f(x). \qquad (1.42)$$

1.3 Main Results

On the way to prove the general Chebyshev–Grüss-type inequalities, we establish the general

Theorem 1.20. *For f, g under the assumptions of any of Theorem 1.10, Corollary 1.11 and Theorem 1.14 we obtain that*

$$\Delta(f, g) := \left| \int_\alpha^\beta \omega(x) f(x) g(x) \, dx - \left(\int_\alpha^\beta \omega(x) f(x) \, dx \right) \left(\int_\alpha^\beta \omega(x) g(x) \, dx \right) \right|$$

$$\leq \frac{1}{2} \left[\left(\int_\alpha^\beta |\omega(x)| \, |g(x)| \, |Q^{l-1} f(x)| \, dx + \int_\alpha^\beta |\omega(x)| \, |f(x)| \, |Q^{l-1} g(x)| \, dx \right) \right.$$

$$\left. + \left(\int_\alpha^\beta |\omega(x)| \, |g(x)| \, |R_{0,l} f(x)| \, dx + \int_\alpha^\beta |\omega(x)| \, |f(x)| \, |R_{0,l} g(x)| \, dx \right) \right]. \qquad (1.43)$$

Proof. For $x \in (\alpha, \beta)$ we have

$$f(x) = \int_\alpha^\beta f(y) \omega(y) \, dy + Q^{l-1} f(x) + R_{0,l} f(x),$$

and

$$g(x) = \int_\alpha^\beta g(y) \omega(y) \, dy + Q^{l-1} g(x) + R_{0,l} g(x).$$

Hence

$$\omega(x) f(x) g(x) = \omega(x) g(x) \int_\alpha^\beta f(y) \omega(y) \, dy$$

$$+ \omega(x) g(x) Q^{l-1} f(x) + \omega(x) g(x) R_{0,l} f(x),$$

and

$$\omega(x) f(x) g(x) = \omega(x) f(x) \int_\alpha^\beta g(y) \omega(y) \, dy$$
$$+ \omega(x) f(x) Q^{l-1} g(x) + \omega(x) f(x) R_{0,l} g(x).$$

Therefore

$$\int_\alpha^\beta \omega(x) f(x) g(x) \, dx = \left(\int_\alpha^\beta \omega(x) g(x) \, dx \right) \left(\int_\alpha^\beta f(x) \omega(x) \, dx \right)$$
$$+ \int_\alpha^\beta \omega(x) g(x) Q^{l-1} f(x) \, dx$$
$$+ \int_\alpha^\beta \omega(x) g(x) R_{0,l} f(x) \, dx,$$

and

$$\int_\alpha^\beta \omega(x) f(x) g(x) \, dx = \left(\int_\alpha^\beta \omega(x) f(x) \, dx \right) \left(\int_\alpha^\beta g(x) \omega(x) \, dx \right)$$
$$+ \int_\alpha^\beta \omega(x) f(x) Q^{l-1} g(x) \, dx$$
$$+ \int_\alpha^\beta \omega(x) f(x) R_{0,l} g(x) \, dx.$$

Consequently it holds

$$\int_\alpha^\beta \omega(x) f(x) g(x) \, dx - \left(\int_\alpha^\beta \omega(x) f(x) \, dx \right) \left(\int_\alpha^\beta g(x) \omega(x) \, dx \right)$$
$$= \int_\alpha^\beta \omega(x) g(x) Q^{l-1} f(x) \, dx + \int_a^\beta \omega(x) g(x) R_{0,l} f(x) \, dx,$$

and

$$\int_\alpha^\beta \omega(x) f(x) g(x) \, dx - \left(\int_\alpha^\beta \omega(x) f(x) \, dx \right) \left(\int_\alpha^\beta g(x) \omega(x) \, dx \right)$$
$$= \int_\alpha^\beta \omega(x) f(x) Q^{l-1} g(x) \, dx + \int_\alpha^\beta \omega(x) f(x) R_{0,l} g(x) \, dx.$$

Adding the last two equalities and dividing by two, we get

$$\int_{\alpha}^{\beta} \omega(x) f(x) g(x) dx - \left(\int_{\alpha}^{\beta} \omega(x) f(x) dx \right) \left(\int_{\alpha}^{\beta} g(x) \omega(x) dx \right)$$

$$= \frac{1}{2} \left[\left(\int_{\alpha}^{\beta} \omega(x) g(x) Q^{l-1} f(x) dx + \int_{\alpha}^{\beta} \omega(x) f(x) Q^{l-1} g(x) dx \right) \right.$$

$$\left. + \left(\int_{\alpha}^{\beta} \omega(x) g(x) R_{0,l} f(x) dx + \int_{\alpha}^{\beta} \omega(x) f(x) R_{0,l} g(x) dx \right) \right],$$

hence proving the claim. ∎

General Chebyshev–Grüss inequalities follow.
We give

Theorem 1.21. *Let f, g with $f^{(l-1)}$, $g^{(l-1)}$ absolutely continuous on $[a, b] \subset \mathbb{R}$, $l \in \mathbb{N}$; $(\alpha, \beta) \subset (a, b)$. Let also $\omega \in L_1(\mathbb{R})$, $\sup p\omega \subset [\alpha, \beta]$, $\int_{\mathbb{R}} \omega(x) dx = 1$. Then*

$$\left| \int_{\alpha}^{\beta} \omega(x) f(x) g(x) dx - \left(\int_{\alpha}^{\beta} \omega(x) f(x) dx \right) \left(\int_{\alpha}^{\beta} \omega(x) g(x) dx \right) \right|$$

$$\leq \frac{\|\omega\|_{L_1(\mathbb{R})}^2}{2} \left[\left[\|g\|_{\infty,(\alpha,\beta)} \left(\sum_{k=1}^{l-1} \frac{\|f^{(k)}\|_{\infty,(\alpha,\beta)} (\beta - \alpha)^k}{k!} \right) \right. \right.$$

$$\left. + \|f\|_{\infty,(\alpha,\beta)} \left(\sum_{k=1}^{l-1} \frac{\|g^{(k)}\|_{\infty,(\alpha,\beta)} (\beta - \alpha)^k}{k!} \right) \right]$$

$$+ \left[\left(\|g\|_{\infty,(\alpha,\beta)} \left\| f^{(l)} \right\|_{L_1(a,\beta)} \right. \right.$$

$$\left. \left. \left. + \|f\|_{\infty,(\alpha,\beta)} \left\| g^{(l)} \right\|_{L_1(\alpha,\beta)} \right) \frac{(\beta - \alpha)^{l-1}}{(l-1)!} \right] \right]. \quad (1.44)$$

Proof. By (1.23) and (1.32). ∎

Theorem 1.22. *Let $f, g \in C^l([a, b])$, $[a, b] \subset \mathbb{R}$, $l \in \mathbb{N}$, $(\alpha, \beta) \subset (a, b)$. Let also $\omega \in L_{\infty}(\mathbb{R})$, $\sup p\omega \subset [\alpha, \beta]$, $\int_{\mathbb{R}} \omega(x) dx = 1$. Then*

$$\left| \int_{\alpha}^{\beta} \omega(x) f(x) g(x) dx - \left(\int_{\alpha}^{\beta} \omega(x) f(x) dx \right) \left(\int_{\alpha}^{\beta} \omega(x) g(x) dx \right) \right|$$

$$\leq \|\omega\|_{L_1(\mathbb{R})} \left[\frac{\|\omega\|_{L_1(\mathbb{R})}}{2} \left\{ \|g\|_{\infty,(\alpha,\beta)} \left(\sum_{k=1}^{l-1} \left(\frac{\|f^{(k)}\|_{\infty,(\alpha,\beta)} (\beta - \alpha)^k}{k!} \right) \right) \right. \right.$$

$$+ \|f\|_{\infty,(\alpha,\beta)} \left(\sum_{k=1}^{l-1} \left(\frac{\|g^{(k)}\|_{\infty,(\alpha,\beta)} (\beta-\alpha)^k}{k!} \right) \right) \Big\}$$

$$+ \left[\|\omega\|_{\infty,(\alpha,\beta)} \frac{(\beta-\alpha)^{l+1}}{(l+1)!} \left(\|g\|_{\infty,(\alpha,\beta)} \|f^{(l)}\|_{\infty,(\alpha,\beta)} \right. \right.$$

$$\left. \left. + \|f\|_{\infty,(\alpha,\beta)} \|g^{(l)}\|_{\infty,(\alpha,\beta)} \right) \right] \Big]. \tag{1.45}$$

Proof. By (1.24) and (1.32). ∎

We further present

Theorem 1.23. *Let* $f, g \in \left(W_1^l\right)^{\text{loc}}(a,b)$; $a,b \in \mathbb{R}$; $(\alpha,\beta) \subset (a,b)$, $l \in \mathbb{N}$; $\omega \in L_\infty(\mathbb{R})$, $\sup p\omega \subset [\alpha,\beta]$, $\int_\mathbb{R} \omega(x)\,dx = 1$. *Then*

$$\left| \int_\alpha^\beta \omega(x) f(x) g(x)\,dx - \left(\int_\alpha^\beta \omega(x) f(x)\,dx \right) \left(\int_\alpha^\beta \omega(x) g(x)\,dx \right) \right|$$

$$\leq \frac{\|\omega\|_{L_\infty(\mathbb{R})}^2}{2} \left\{ \left[\|g\|_{L_1(\alpha,\beta)} \left(\sum_{k=1}^{l-1} \left(\frac{(\beta-\alpha)^k}{k!} \|f_w^{(k)}\|_{L_1(\alpha,\beta)} \right) \right) \right.\right.$$

$$\left. + \|f\|_{L_1(\alpha,\beta)} \left(\sum_{k=1}^{l-1} \left(\frac{(\beta-\alpha)^k}{k!} \|g_w^{(k)}\|_{L_1(\alpha,\beta)} \right) \right) \right]$$

$$+ \left[\left(\|g\|_{L_1(\alpha,\beta)} \|f_w^{(l)}\|_{L_1(\alpha,\beta)} \right.\right.$$

$$\left.\left. + \|f\|_{L_1(\alpha,\beta)} \|g_w^{(l)}\|_{L_1(\alpha,\beta)} \right) \frac{(\beta-\alpha)^l}{(l-1)!} \right] \right\}. \tag{1.46}$$

Proof. By (1.23) and (1.29). ∎

Theorem 1.24. *Let* $f, g \in \left(W_1^l\right)^{\text{loc}}(a,b)$; $a,b \in \mathbb{R}$; $(\alpha,\beta) \subset (a,b)$, $l \in \mathbb{N}$; $\omega \in L_\infty(\mathbb{R})$, $\sup p\omega \subset [\alpha,\beta]$, $\int_\mathbb{R} \omega(x)\,dx = 1$. *Furthermore assume* $f_w^{(k)}, g_w^{(k)} \in L_\infty(\alpha,\beta)$, $k = 1,...,l$. *Then*

$$\left| \int_\alpha^\beta \omega(x) f(x) g(x)\,dx - \left(\int_\alpha^\beta \omega(x) f(x)\,dx \right) \left(\int_\alpha^\beta \omega(x) g(x)\,dx \right) \right|$$

$$\leq \frac{\|\omega\|_{L_\infty(\mathbb{R})}^2}{2} \left\{ \left[\|g\|_{L_1(\alpha,\beta)} \left(\sum_{k=1}^{l-1} \frac{\|f_w^{(k)}\|_{L_\infty(\alpha,\beta)} (\beta-\alpha)^{k+1}}{k!} \right) \right.\right.$$

$$+ \|f\|_{L_1(\alpha,\beta)} \left(\sum_{k=1}^{l-1} \frac{\left\| g_w^{(k)} \right\|_{L_\infty(\alpha,\beta)} (\beta-\alpha)^{k+1}}{k!} \right) \Bigg]$$

$$+ \Bigg[\left(\|g\|_{L_1(\alpha,\beta)} \left\| f_w^{(l)} \right\|_{L_\infty(\alpha,\beta)} \right.$$

$$\left. + \|f\|_{L_1(\alpha,\beta)} \left\| g_w^{(l)} \right\|_{L_\infty(\alpha,\beta)} \right) \frac{(\beta-\alpha)^{l+1}}{(l-1)!} \Bigg] \Bigg\}. \quad (1.47)$$

Proof. As in (1.24) and by (1.30). ∎

Theorem 1.25. *Let* $f, g \in \left(W_1^l \right)^{loc} (a,b)$; $a, b \in \mathbb{R}$; $(\alpha, \beta) \subset (a,b)$, $l \in \mathbb{N}$; $\omega \in L_\infty(\mathbb{R})$, $\sup p\omega \subset [\alpha, \beta]$, $\int_{\mathbb{R}} \omega(x)\, dx = 1$. *Furthermore, assume for* $p > 1$ *that* $f_w^{(k)}, g_w^{(k)} \in L_p(\alpha,\beta)$, $k = 1,...,l$. *Then*

$$\Delta(f,g) := \left| \int_\alpha^\beta \omega(x)\, f(x)\, g(x)\, dx - \left(\int_\alpha^\beta \omega(x)\, f(x)\, dx \right) \right.$$

$$\left. \times \left(\int_\alpha^\beta \omega(x)\, g(x)\, dx \right) \right|$$

$$\leq \frac{\|\omega\|_{L_\infty(\mathbb{R})}^2}{2} \Bigg[\Bigg\{ \|g\|_{L_1(\alpha,\beta)} \left(\sum_{k=1}^{l-1} \frac{(\beta-\alpha)^{k+1-\frac{1}{p}}}{k!} \left\| f_w^{(k)} \right\|_{L_p(\alpha,\beta)} \right)$$

$$+ \|f\|_{L_1(\alpha,\beta)} \left(\sum_{k=1}^{l-1} \frac{(\beta-\alpha)^{k+1-\frac{1}{p}}}{k!} \left\| g_w^{(k)} \right\|_{L_p(\alpha,\beta)} \right) \Bigg\}$$

$$+ \Bigg\{ \left(\|g\|_{L_1(\alpha,\beta)} \left\| f_w^{(l)} \right\|_{L_p(\alpha,\beta)} \right.$$

$$\left. + \|f\|_{L_1(\alpha,\beta)} \left\| g_w^{(l)} \right\|_{L_p(\alpha,\beta)} \right) \frac{(\beta-\alpha)^{l+1-\frac{1}{p}}}{(l-1)!} \Bigg\} \Bigg].$$

$$(1.48)$$

Proof. Working as in (1.25) and from (1.33). ∎

Remark 1.26. When $f, g \in C^l([a,b])$, $l \in \mathbb{N}$, Theorems 1.23–1.25 are again valid. In this case, we substitute $f_w^{(k)}$, $g_w^{(k)}$ by $f^{(k)}$, $g^{(k)}$ in all inequalities (1.46)–(1.48); $k = 1,...,l$.

We continue with

Theorem 1.27. *Let* $l \in \mathbb{N} - \{1\}$, $\omega \in C^{(l-2)}(\mathbb{R})$, $\sup p\omega \subset [\alpha, \beta]$, $\int_{\mathbb{R}} \omega(x)\, dx = 1$, *and the derivative* $\omega^{(l-2)}$ *is absolutely continuous on* $[a,b] \subset \mathbb{R}$; $(\alpha, \beta) \subset (a,b)$. *Here suppose* $f, g \in .\left(W_1^l \right)^{loc}(a,b)$, *or* $f, g \in C^l([a,b])$. *Here* $\overline{f}^{(l)}$ *denotes either* $f_w^{(l)}$ *or* $f^{(l)}$, *and* $\Delta(f,g)$ *as in (1.43).*

We have the following cases:

(i) It holds

$$\Delta(f,g) \leq \frac{\|\omega\|_\infty}{2} \left[2\,\|g\|_{L_1(\alpha,\beta)}\,\|f\|_{L_1(\alpha,\beta)} \right.$$

$$\times \left(\sum_{k=1}^{l-1} \frac{1}{k!} \sup_{x\in(\alpha,\beta)} \left\| \left[(x-y)^k\,\omega(y) \right]_y^{(k)} \right\|_\infty \right)$$

$$+ \left(\|g\|_{L_1(\alpha,\beta)} \left\| \overline{f}^{(l)} \right\|_{L_1(\alpha,\beta)} \right.$$

$$\left. \left. + \|f\|_{L_1(\alpha,\beta)} \left\| \overline{g}^{(l)} \right\|_{L_1(\alpha,\beta)} \right) \frac{(\beta-\alpha)^l}{(l-1)!}\,\|\omega\|_\infty \right]. \quad (1.49)$$

(ii) Assume further that $f, g, \overline{f}^{(l)}, \overline{g}^{(l)} \in L_\infty(\alpha,\beta)$. Then

$$\Delta(f,g) \leq \frac{\|\omega\|_\infty}{2} \left[\left\{ \left(\|f\|_{L_\infty(\alpha,\beta)}\,\|g\|_{L_1(\alpha,\beta)} + \|g\|_{L_\infty(\alpha,\beta)}\,\|f\|_{L_1(\alpha,\beta)} \right) \right. \right.$$

$$\times \left(\sum_{k=1}^{l-1} \frac{1}{k!} \sup_{x\in(\alpha,\beta)} \left\| \left[(x-y)^k\,\omega(y) \right]_y^{(k)} \right\|_{L_1(\alpha,\beta)} \right) \right\}$$

$$+ \left\{ \left(\|g\|_{L_1(\alpha,\beta)} \left\| \overline{f}^{(l)} \right\|_{L_\infty(\alpha,\beta)} \right. \right.$$

$$\left. \left. \left. + \|f\|_{L_1(\alpha,\beta)} \left\| \overline{g}^{(l)} \right\|_{L_\infty(\alpha,\beta)} \right) \frac{(\beta-\alpha)^{l+1}}{(l-1)!}\,\|\omega\|_\infty \right\} \right].$$

$$(1.50)$$

(iii) Let $p,q > 1 : \frac{1}{p} + \frac{1}{q} = 1$; assume further that $f, g, \overline{f}^{(l)}, \overline{g}^{(l)} \in L_p(\alpha,\beta)$. Then

$$\Delta(f,g) \leq \frac{\|\omega\|_\infty}{2} \left[\left\{ \left(\|g\|_{L_1(\alpha,\beta)}\,\|f\|_{L_p(\alpha,\beta)} + \|f\|_{L_1(\alpha,\beta)}\,\|g\|_{L_p(\alpha,\beta)} \right) \right. \right.$$

$$\times \left(\sum_{k=1}^{l-1} \frac{1}{k!} \sup_{x\in(\alpha,\beta)} \left\| \left[(x-y)^k\,\omega(y) \right]_y^{(k)} \right\|_{L_q(\alpha,\beta)} \right) \right\}$$

$$+ \left\{ \left(\|g\|_{L_1(\alpha,\beta)} \left\| \overline{f}^{(l)} \right\|_{L_p(\alpha,\beta)} \right. \right.$$

$$\left. + \|f\|_{L_1(\alpha,\beta)} \left\| \overline{g}^{(l)} \right\|_{L_p(\alpha,\beta)} \right)$$

$$\left. \left. \times \frac{(\beta-\alpha)^{l+1-\frac{1}{p}}}{(l-1)!}\,\|\omega\|_\infty \right\} \right]. \quad (1.51)$$

Proof. By (1.43), (1.37) and by Theorems 1.23, 1.24, 1.25.   ■

Next, we give a series of Ostrowski-type inequalities.

Theorem 1.28. *Let* $l \in \mathbb{N}, [a,b] \subset \mathbb{R}, a < \alpha < \beta < b$ *and* $\omega \in L_1(\mathbb{R})$, $\sup p\omega \subset [\alpha, \beta], \int_{\mathbb{R}} \omega(x) \, dx = 1$. *Assume* f *on* $[a,b]$: $f^{(l-1)}$ *exists and is absolutely continuous on* $[a,b]$. *Then for any* $x \in (\alpha, \beta)$ *we get*

$$\left| f(x) - \int_{\alpha}^{\beta} f(y) \omega(y) \, dy - Q^{l-1} f(x) \right|$$

$$\leq \frac{\|\omega\|_{L_1(\mathbb{R})} \|f^{(l)}\|_{L_1(\alpha,\beta)} (\beta - \alpha)^{l-1}}{(l-1)!} := A_1. \tag{1.52}$$

If additionally we assume $\|\omega\|_{L_\infty(\mathbb{R})} < \infty$, *then* $\forall \ x \in (\alpha, \beta)$, *we obtain*

$$\left| f(x) - \int_{\alpha}^{\beta} f(y) \omega(y) \, dy \right|$$

$$\leq \left(\sum_{k=1}^{l-1} \left(\frac{(\beta - x)^{k+1} + (x - \alpha)^{k+1}}{(k+1)!} \right) \|f^{(k)}\|_{\infty} \right) \|\omega\|_{L_\infty(\mathbb{R})}$$

$$+ \frac{\|\omega\|_{L_1(\mathbb{R})} \|f^{(l)}\|_{L_1(\alpha,\beta)} (\beta - \alpha)^{l-1}}{(l-1)!} := B_1(x). \tag{1.53}$$

Proof. By (1.41), (1.23), and (1.30).   ■

Theorem 1.29. *All as in Theorem 1.28. Assume* $f \in C^l([a,b])$. *Then* $\forall \ x \in (\alpha, \beta)$,

$$\left| f(x) - \int_{\alpha}^{\beta} f(y) \omega(y) \, dy - Q^{l-1} f(x) \right|$$

$$\leq \frac{\|\omega\|_{L_1(\mathbb{R})} \|f^{(l)}\|_{\infty} \left((\beta - x)^l + (x - \alpha)^l \right)}{l!} =: A_2(x). \tag{1.54}$$

If additionally we assume $\|\omega\|_{L_\infty(\mathbb{R})} < \infty$, *then* $\forall \ x \in (\alpha, \beta)$,

$$\left| f(x) - \int_{\alpha}^{\beta} f(y) \omega(y) \, dy \right|$$

$$\leq \left(\sum_{k=1}^{l-1} \left(\frac{(\beta - x)^{k+1} + (x - \alpha)^{k+1}}{(k+1)!} \right) \|f^{(k)}\|_{\infty} \right) \|\omega\|_{L_\infty(\mathbb{R})}$$

$$+ \frac{\|\omega\|_{L_1(\mathbb{R})} \|f^{(l)}\|_{\infty} \left((\beta - x)^l + (x - \alpha)^l \right)}{l!} =: B_2(x). \tag{1.55}$$

Proof. By (1.41), (1.24), and (1.30). ∎

We continue with

Theorem 1.30. *Let all as in Theorem 1.28 or* $f \in \left(W_1^l\right)^{loc}(a,b)$ *and rest as in Theorem 1.28. Then* $\forall\ x \in (\alpha, \beta)$ *(or almost every* $x \in (\alpha, \beta)$*, respectively), we get*

$$E(f)(x) := \left| f(x) - \int_\alpha^\beta f(y)\,\omega(y)\,\mathrm{d}y - Q^{l-1}f(x) \right|$$

$$\leq \frac{\|\omega\|_{L_1(a,b)} \left\|\overline{f}^{(l)}\right\|_{L_1(\alpha,\beta)} (\beta-\alpha)^{l-1}}{(l-1)!} =: A_3 \qquad (1.56)$$

Additionally, if $\|\omega\|_{L_\infty(\mathbb{R})} < \infty$*,* $\forall\ x \in (\alpha, \beta)$ *(or almost every* $x \in (\alpha, \beta)$*, respectively), we derive*

$$\Delta(f)(x) := \left| f(x) - \int_\alpha^\beta f(y)\,\omega(y)\,\mathrm{d}y \right|$$

$$\leq \left(\sum_{k=1}^{l-1} \frac{(\beta-\alpha)^k}{k!} \left\|\overline{f}^{(k)}\right\|_{L_1(\alpha,\beta)} \right) \|\omega\|_{L_\infty(\mathbb{R})}$$

$$+ \frac{\|\omega\|_{L_1(a,b)} \left\|\overline{f}^{(l)}\right\|_{L_1(\alpha,\beta)} (\beta-\alpha)^{l-1}}{(l-1)!} =: B_3. \qquad (1.57)$$

Proof. By (1.41), (1.23), and (1.29). ∎

Theorem 1.31. *Let all as in Theorem 1.28 or* $f \in \left(W_1^l\right)^{loc}(a,b)$ *and rest as in Theorem 1.28. Assume further* $\overline{f}^{(l)} \in L_\infty(\alpha, \beta)$*. Then* $\forall\ x \in (\alpha, \beta)$ *(or almost every* $x \in (\alpha, \beta)$*, respectively), we get*

$$E(f)(x) \leq \frac{\|\omega\|_{L_1(\mathbb{R})} \left\|\overline{f}^{(l)}\right\|_{L_\infty(\alpha,\beta)} \left((\beta-x)^l + (x-\alpha)^l\right)}{l!} =: A_4(x).$$

$$(1.58)$$

Additionally $\overline{f}^{(k)} \in L_\infty(\alpha, \beta)$, $k = 1, ..., l - 1$ and if $\|\omega\|_{L_\infty(\mathbb{R})} < \infty$, then $\forall$ $x \in (\alpha, \beta)$ (or almost every $x \in (\alpha, \beta)$, respectively), we get

$$\Delta(f)(x) \leq \left(\sum_{k=1}^{l-1} \left(\frac{\left((\beta - x)^{k+1} + (x - \alpha)^{k+1} \right)}{(k + 1)!} \right) \left\| \overline{f}^{(k)} \right\|_{L_\infty(\alpha, \beta)} \right) \|\omega\|_{L_\infty(\mathbb{R})}$$

$$+ \frac{\|\omega\|_{L_1(\mathbb{R})} \left\| \overline{f}^{(l)} \right\|_{L_\infty(\alpha, \beta)} \left((\beta - x)^l + (x - \alpha)^l \right)}{l!}$$

$$=: B_4(x). \tag{1.59}$$

Proof. By (1.41), (1.24), and (1.30). ∎

Theorem 1.32. *Let all as in Theorem 1.28 or $f \in \left(W_1^l \right)^{\mathrm{loc}}(a, b)$ and rest as in Theorem 1.28. Let $p, q > 1 : \frac{1}{p} + \frac{1}{q} = 1$. Assume further $\overline{f}^{(l)} \in L_p(\alpha, \beta)$. Then $\forall\, x \in (\alpha, \beta)$ (or almost every $x \in (\alpha, \beta)$, respectively), we get*

$$E(f)(x) \leq \frac{\|\omega\|_{L_1(\mathbb{R})} \left\| \overline{f}^{(l)} \right\|_{L_p(\alpha, \beta)}}{(l - 1)!}$$

$$\times \left(\frac{(\beta - x)^{q(l-1)+1} + (x - \alpha)^{q(l-1)+1}}{q(l - 1) + 1} \right)^{\frac{1}{q}} =: A_5(x). \tag{1.60}$$

Additionally, if $\overline{f}^{(k)} \in L_p(\alpha, \beta)$, $k = 1, ..., l - 1$ and $\|\omega\|_{L_\infty(\mathbb{R})} < \infty$, then $\forall$ $x \in (\alpha, \beta)$ (or almost every $x \in (\alpha, \beta)$, respectively), we get

$$\Delta(f)(x)$$

$$\leq \left(\sum_{k=1}^{l-1} \left(\frac{\left((\beta - x)^{(kq+1)} + (x - \alpha)^{(kq+1)} \right)}{kq + 1} \right)^{\frac{1}{q}} \frac{\left\| \overline{f}^{(k)} \right\|_{L_p(\alpha, \beta)}}{k!} \right)$$

$$\times \|\omega\|_{L_\infty(\mathbb{R})} + A_5(x) =: B_5(x). \tag{1.61}$$

Proof. By (1.41), (1.25), and (1.31). ∎

We further give

Theorem 1.33. *Let all as in Theorem 1.31. Here assume* $\omega \in L_1(\mathbb{R})$. *Then* $\forall$ $x \in (\alpha, \beta)$ *(or almost every* $x \in (\alpha, \beta)$, *respectively), we get*

$$\Delta(f)(x) := \left| f(x) - \int_{\alpha}^{\beta} f(y)\,\omega(y)\,\mathrm{d}y \right|$$

$$\leq \left(\sum_{k=1}^{l-1} \frac{\left\| \overline{f}^{(k)} \right\|_{L_{\infty}(\alpha,\beta)} (\beta-\alpha)^k}{k!} \right) \|\omega\|_{L_1(\mathbb{R})} + A_4(x)$$

$$=: B_6(x). \tag{1.62}$$

Proof. By (1.32) and (1.58). ∎

Theorem 1.34. *Let all be as in Theorem 1.32. Here assume* $\omega \in L_q(\alpha, \beta)$, $q > 1$. *Then* $\forall$ $x \in (\alpha, \beta)$ *(or almost every* $x \in (\alpha, \beta)$, *respectively), it holds*

$$\Delta(f)(x) \leq \left(\sum_{k=1}^{l-1} \frac{(\beta-\alpha)^k}{k!} \left\| \overline{f}^{(k)} \right\|_{L_p(\alpha,\beta)} \right) \|\omega\|_{L_q(\alpha,\beta)} + A_5(x) =: B_7(x).$$

$$\tag{1.63}$$

Proof. By (1.33) and (1.60). ∎

Theorem 1.35. *Let all as in Theorem 1.28 or* $f \in \left(W_1^l\right)^{\mathrm{loc}}(a,b)$ *and rest as in Theorem 1.28. Let* $p, q, r > 1 : \frac{1}{p} + \frac{1}{q} + \frac{1}{r} = 1$, $\overline{f}^{(k)} \in L_p(\alpha, \beta)$, $k = 1, ..., l-1$; $\omega \in L_q(\alpha, \beta)$. *Then* $\forall$ $x \in (\alpha, \beta)$ *(or almost every* $x \in (\alpha, \beta)$, *respectively), it holds*

$$\left| f(x) - \int_{\alpha}^{\beta} f(y)\,\omega(y)\,\mathrm{d}y - R_{0,l}f(x) \right|$$

$$\leq \left(\sum_{k=1}^{l-1} \frac{\left\| \overline{f}^{(k)} \right\|_{L_p(\alpha,\beta)}}{k!} \left(\frac{(\beta-x)^{(kr+1)} + (x-\alpha)^{(kr+1)}}{(kr+1)} \right)^{\frac{1}{r}} \right) \|\omega\|_{L_q(\alpha,\beta)}$$

$$=: \Phi(x). \tag{1.64}$$

Proof. By (1.34) and (1.41). ∎

We also give

Theorem 1.36. *Let* $\mathbb{N} \ni l > 1$ *and* $\omega \in C^{(l-2)}(\mathbb{R})$, $\sup p\omega \subset [\alpha, \beta]$, $\int_{\mathbb{R}} \omega(x)\,\mathrm{d}x = 1$, $\omega^{(l-2)}$ *is absolutely continuous on* $[a, b]$, $[\alpha, \beta] \subset (a, b)$; $a, b \in \mathbb{R}$. *Here* $f \in C^l([a,b])$ *or* $f \in \left(W_1^l\right)^{\mathrm{loc}}(a,b)$. *For every* $x \in (\alpha, \beta)$ *(or almost every* $x \in (\alpha, \beta)$, *respectively), we get for*

$$\Delta(f)(x) := \left| f(x) - \int_{\alpha}^{\beta} f(y)\,\omega(y)\,\mathrm{d}y \right|$$

that:

(i) It holds

$$\Delta(f)(x) \le \left(\sum_{k=1}^{l-1} \frac{1}{k!} \left\| \left[(x-y)^k \omega(y)\right]_y^{(k)} \right\|_\infty \right) \|f\|_{L_1(\alpha,\beta)}$$

$$+ \frac{\|\omega\|_{L_1(\mathbb{R})} \left\| \overline{f}^{(l)} \right\|_{L_1(\alpha,\beta)} (\beta-\alpha)^{l-1}}{(l-1)!} =: C_1(x). \quad (1.65)$$

(ii) If $f, \overline{f}^{(l)} \in L_\infty(\alpha,\beta)$, then

$$\Delta(f)(x) \le \left(\sum_{k=1}^{l-1} \frac{1}{k!} \left\| \left[(x-y)^k \omega(y)\right]_y^{(k)} \right\|_{L_1(\alpha,\beta)} \right) \|f\|_{L_\infty(\alpha,\beta)}$$

$$+ \frac{\|\omega\|_{L_1(\mathbb{R})} \left\| \overline{f}^{(l)} \right\|_{L_\infty(\alpha,\beta)} \left((\beta-x)^l + (x-\alpha)^l\right)}{l!}$$

$$=: C_2(x). \quad (1.66)$$

(iii) Let $p,q > 1 : \frac{1}{p} + \frac{1}{q} = 1$. Assume further that $f, \overline{f}^{(l)} \in L_p(\alpha,\beta)$. Then

$$\Delta(f)(x) \le \left(\sum_{k=1}^{l-1} \frac{1}{k!} \left\| \left[(x-y)^k \omega(y)\right]_y^{(k)} \right\|_{L_q(\alpha,\beta)} \right) \|f\|_{L_p(\alpha,\beta)}$$

$$+ \frac{\|\omega\|_{L_1(\mathbb{R})} \left\| \overline{f}^{(l)} \right\|_{L_p(\alpha,\beta)}}{(l-1)!}$$

$$\times \left(\frac{(\beta-x)^{q(l-1)+1} + (x-\alpha)^{q(l-1)+1}}{q(l-1)+1} \right)^{\frac{1}{q}}$$

$$=: C_3(x). \quad (1.67)$$

Proof. By (1.37) and Theorems 1.30–1.32. ∎

We finish Ostrowski-type inequalities with

Theorem 1.37. *Let $l, m \in \mathbb{N}$, $m < l$; $\omega \in C^{(l-2)}(\mathbb{R})$, sup $p\omega \subset [\alpha,\beta]$, $\int_\mathbb{R} \omega(x)\,dx = 1$, $\omega^{(l-2)}$ is absolutely continuous on $[a,b]$, $[\alpha,\beta] \subset (a,b)$; $a, b \in \mathbb{R}$. Here $f \in C^l([a,b])$ or $f \in \left(W_1^l\right)^{loc}(a,b)$. For every $x \in (\alpha,\beta)$ (or almost every $x \in (\alpha,\beta)$, respectively), we get for*

$$E_\beta(f)(x) := \left| \overline{f}^{(m)}(x) - (-1)^m \int_\alpha^\beta f(y)\omega^{(m)}(y)\,dy - Q_m^{l-1}f(x) \right|, \quad (1.68)$$

and

$$\Delta_\beta (f)(x) := \left| \overline{f}^{(m)}(x) - (-1)^m \int_\alpha^\beta f(y)\,\omega^{(m)}(y)\,\mathrm{d}y \right| \qquad (1.69)$$

that

(i) It holds

$$E_\beta (f)(x) \le \frac{\|\omega\|_{L_1(\mathbb{R})} \left\| \overline{f}^{(l)} \right\|_{L_1(\alpha,\beta)} (\beta - \alpha)^{l-m-1}}{(l-m-1)!} =: E_1, \qquad (1.70)$$

and

$$\Delta_\beta (f)(x) \le \left(\sum_{k=1}^{l-m-1} \frac{1}{k!} \left\| \left[(x-y)^k\,\omega(y) \right]_y^{(k+m)} \right\|_\infty \right) \|f\|_{L_1(\alpha,\beta)}$$

$$+ E_1 =: G_1(x). \qquad (1.71)$$

(ii) If $\overline{f}^{(l)} \in L_\infty (\alpha, \beta)$, then

$$E_\beta (f)(x) \le \frac{\|\omega\|_{L_1(\mathbb{R})} \left\| \overline{f}^{(l)} \right\|_{L_\infty(\alpha,\beta)}}{(l-m)!}$$

$$\times \left((\beta - x)^{l-m} + (x-\alpha)^{l-m} \right) =: E_2(x), \qquad (1.72)$$

if additionally we assume $f \in L_\infty (\alpha, \beta)$, then

$$\Delta_\beta (f)(x) \le \left(\sum_{k=1}^{l-m-1} \frac{1}{k!} \left\| \left[(x-y)^k\,\omega(y) \right]_y^{(k)} \right\|_{L_1(\alpha,\beta)} \right)$$

$$\times \|f\|_{L_\infty(\alpha,\beta)} + E_2(x) =: G_2(x). \qquad (1.73)$$

(iii) Let $p,q > 1 : \frac{1}{p} + \frac{1}{q} = 1$, assume further that $\overline{f}^{(l)} \in L_p (\alpha, \beta)$, then

$$E_\beta (f)(x) \le \frac{\|\omega\|_{L_1(\mathbb{R})} \left\| \overline{f}^{(l)} \right\|_{L_p(\alpha,\beta)}}{(l-m-1)!}$$

$$\times \left(\frac{(\beta - x)^{q(l-m-1)+1} + (x-\alpha)^{q(l-m-1)+1}}{q(l-m-1)+1} \right)^{\frac{1}{q}}$$

$$=: E_3(x), \qquad (1.74)$$

and if additionally $f \in L_p(\alpha, \beta)$, then

$$\Delta_\beta(f)(x) \leq \left(\sum_{k=1}^{l-m-1} \frac{1}{k!} \left\| \left[(x-y)^k \, \omega(y) \right]_y^{(k+m)} \right\|_{L_q(\alpha,\beta)} \right) \| f \|_{L_p(\alpha,\beta)}$$

$$+ E_3(x) =: G_3(x). \tag{1.75}$$

Proof. By (1.23)–(1.25), (1.40), and (1.42). ∎

We make

Remark 1.38. In preparation to present comparison of integral means inequalities, we consider $(\alpha_1, \beta_1) \subseteq (\alpha, \beta)$. We consider also a weight function $\psi \geq 0$ which is Lebesgue integrable on $\mathbb{R}$ with sup $p\psi \subset [\alpha_1, \beta_1] \subset [a,b]$, and $\int_{\mathbb{R}} \psi(x)\,dx = 1$. Clearly here $\int_{\alpha_1}^{\beta_1} \psi(x)\,dx = 1$.

For example, for $x \in (\alpha_1, \beta_1)$, $\psi(x) := \frac{1}{\beta_1 - \alpha_1}$, zero elsewhere, etc.

We will apply the following principle: In general a constraint of the form $|F(x) - G| \leq \varepsilon$, where F is a function and G, ε real numbers so that all make sense, implies that

$$\left| \int_{\mathbb{R}} F(x) \, \psi(x)\,dx - G \right| \leq \varepsilon. \tag{1.76}$$

Next we present a series of comparison of integral means inequalities based on Ostrowski-type inequalities given in this chapter. We use Remark 1.38.

Theorem 1.39. *All as in Theorem 1.28. Then*

$$u(f) := \left| \int_{\alpha_1}^{\beta_1} f(x) \, \psi(x)\,dx - \int_{\alpha}^{\beta} f(y) \, \omega(y)\,dy - \int_{\alpha_1}^{\beta_1} Q^{l-1} f(x) \, \psi(x)\,dx \right|$$

$$\leq A_1, \tag{1.77}$$

and

$$m(f) := \left| \int_{\alpha_1}^{\beta_1} f(x) \, \psi(x)\,dx - \int_{\alpha}^{\beta} f(y) \, \omega(y)\,dy \right|$$

$$\leq \left(\sum_{k=1}^{l-1} \frac{(\beta - \alpha)^{k+1}}{(k+1)!} \left\| f^{(k)} \right\|_\infty \right) \| \omega \|_{L_\infty(\mathbb{R})}$$

$$+ \frac{\| \omega \|_{L_1(\mathbb{R})} \left\| f^{(l)} \right\|_{L_1(\alpha,\beta)} (\beta - \alpha)^{l-1}}{(l-1)!}. \tag{1.78}$$

Proof. By Remark 1.38, Theorem 1.28, and the fact that the functions $(\beta - x)^{k+1}$ $+ (x - \alpha)^{k+1}$, $k = 1, ..., l - 1$ are positive and convex with maximum $(\beta - \alpha)^{k+1}$. ∎

Theorem 1.40. *All as in Theorem 1.29. Then*

$$u(f) \leq \frac{\|\omega\|_{L_1(\mathbb{R})} \|f^{(l)}\|_\infty (\beta - \alpha)^l}{l!}, \tag{1.79}$$

and

$$m(f) \leq \left(\sum_{k=1}^{l-1} \frac{(\beta - \alpha)^{k+1}}{(k+1)!} \|f^{(k)}\|_\infty \right) \|\omega\|_{L_\infty(\mathbb{R})}$$

$$+ \frac{\|\omega\|_{L_1(\mathbb{R})} \|f^{(l)}\|_\infty (\beta - \alpha)^l}{l!}. \tag{1.80}$$

Proof. Just maximize $A_2(x)$ of (1.54) and $B_2(x)$ of (1.55), etc. ∎

Theorem 1.41. *All as in Theorem 1.30. Then*

$$u(f) \leq A_3, \tag{1.81}$$

and

$$m(f) \leq B_3. \tag{1.82}$$

Theorem 1.42. *All as in Theorem 1.31. Then*

$$u(f) \leq \frac{\|\omega\|_{L_1(\mathbb{R})} \left\|\overline{f}^{(l)}\right\|_{L_\infty(\alpha,\beta)} (\beta - \alpha)^l}{l!}, \tag{1.83}$$

and

$$m(f) \leq \left(\sum_{k=1}^{l-1} \frac{(\beta - \alpha)^{k+1}}{(k+1)!} \left\|\overline{f}^{(k)}\right\|_{L_\infty(\alpha,\beta)} \right) \|\omega\|_{L_\infty(\mathbb{R})}$$

$$+ \frac{\|\omega\|_{L_1(\mathbb{R})} \left\|\overline{f}^{(l)}\right\|_{L_\infty(\alpha,\beta)} (\beta - \alpha)^l}{l!}. \tag{1.84}$$

Theorem 1.43. *All as in Theorem 1.32. Then*

$$u(f) \leq \int_{\alpha_1}^{\beta_1} A_5(x)\, \psi(x)\, \mathrm{d}x, \tag{1.85}$$

and

$$m(f) \le \int_{\alpha_1}^{\beta_1} B_5(x) \psi(x) \, dx. \tag{1.86}$$

Proof. By the principle: if $|F(x) - G| \le \varepsilon(x)$, then $\left| \int F(x) \psi(x) \, dx - G \right| \le \int \varepsilon(x) \psi(x) \, dx$, etc. Here $A_5(x)$ as in (1.60) and $B_5(x)$ as in (1.61). ■

Theorem 1.44. *All as in Theorem 1.33. Then*

$$m(f) \le \int_{\alpha_1}^{\beta_1} B_6(x) \psi(x) \, dx, \tag{1.87}$$

where $B_6(x)$ as in (1.62).

Theorem 1.45. *All as in Theorem 1.34. Then*

$$m(f) \le \int_{\alpha_1}^{\beta_1} B_7(x) \psi(x) \, dx, \tag{1.88}$$

where $B_7(x)$ as in (1.63).

Theorem 1.46. *All as in Theorem 1.35. Then*

$$\left| \int_{\alpha_1}^{\beta_1} f(x) \psi(x) \, dx - \int_{\alpha}^{\beta} f(y) \omega(y) \, dy - \int_{\alpha_1}^{\beta_1} (R_{0,l} f(x)) \psi(x) \, dx \right|$$

$$\le \int_{\alpha_1}^{\beta_1} \Phi(x) \psi(x) \, dx, \tag{1.89}$$

where $\Phi(x)$ as in (1.64).

We continue with

Theorem 1.47. *All as in Theorem 1.36. Then*

(i)

$$m(f) \le \left(\sum_{k=1}^{l-1} \frac{1}{k!} \sup_{x \in [\alpha_1, \beta_1]} \left\| \left[(x-y)^k \omega(y) \right]_y^{(k)} \right\|_\infty \right) \|f\|_{L_1(\alpha,\beta)}$$

$$+ \frac{\|\omega\|_{L_1(\mathbb{R})} \left\| \overline{f}^{(l)} \right\|_{L_1(\alpha,\beta)} (\beta - \alpha)^{l-1}}{(l-1)!}. \tag{1.90}$$

(ii) If $f, \overline{f}^{(l)} \in L_\infty (\alpha, \beta)$, then

$$m (f) \leq \int_{\alpha_1}^{\beta_1} C_2 (x) \psi (x) \, dx. \tag{1.91}$$

(iii) Let $p, q > 1 : \frac{1}{p} + \frac{1}{q} = 1$; assume further $f, \overline{f}^{(l)} \in L_p (\alpha, \beta)$, then

$$m (f) \leq \int_{\alpha_1}^{\beta_1} C_3 (x) \psi (x) \, dx. \tag{1.92}$$

Here $C_2 (x)$ as in (1.66) and $C_3 (x)$ as in (1.67).

We finish the results about comparison of integral means with

Theorem 1.48. *All as in Theorem 1.37. Denote by*

$$u_m (f) := \left| \int_{\alpha_1}^{\beta_1} \overline{f}^{(m)} (x) \psi (x) \, dx - (-1)^m \int_\alpha^\beta f (y) \omega^{(m)} (y) \, dy \right.$$

$$\left. - \int_{\alpha_1}^{\beta_1} \left(Q_m^{l-1} f (x) \right) \psi (x) \, dx \right|, \tag{1.93}$$

and

$$\rho_m (f) := \left| \int_{\alpha_1}^{\beta_1} \overline{f}^{(m)} (x) \psi (x) \, dx - (-1)^m \int_\alpha^\beta f (y) \omega^{(m)} (y) \, dy \right|. \tag{1.94}$$

(i) It holds

$$u_m (f) \leq E_1, \tag{1.95}$$

where E_1 as in (1.70), and

$$\rho_m (f) \leq \left(\sum_{k=1}^{l-m-1} \frac{1}{k!} \sup_{x \in [\alpha_1, \beta_1]} \left\| \left[(x - y)^k \omega (y) \right]_y^{(k+m)} \right\|_\infty \right) \| f \|_{L_1 (\alpha, \beta)} + E_1, \tag{1.96}$$

(ii) If $\overline{f}^{(l)} \in L_\infty (\alpha, \beta)$, then

$$u_m (f) \leq \frac{\| \omega \|_{L_1 (\mathbb{R})} \left\| \overline{f}^{(l)} \right\|_{L_\infty (\alpha, \beta)}}{(l - m)!} (\beta - \alpha)^{l-m}, \tag{1.97}$$

and if additionally assume $f \in L_\infty(\alpha, \beta)$, *then*

$$\rho_m(f) \le \left(\sum_{k=1}^{l-m-1} \frac{1}{k!} \sup_{x \in [\alpha_1, \beta_1]} \left\| \left[(x-y)^k \omega(y) \right]_y^{(k+m)} \right\|_{L_1(\alpha, \beta)} \right) \|f\|_{L_\infty(\alpha, \beta)}$$

$$+ \frac{\|\omega\|_{L_1(\mathbb{R})} \left\| \overline{f}^{(l)} \right\|_{L_\infty(\alpha, \beta)} (\beta - \alpha)^{l-m}}{(l-m)!}, \tag{1.98}$$

(iii) Let $p, q > 1 : \frac{1}{p} + \frac{1}{q} = 1$, *assume further* $\overline{f}^{(l)} \in L_p(\alpha, \beta)$, *then*

$$u_m(f) \le \int_{\alpha_1}^{\beta_1} E_3(x) \psi(x) \, dx, \tag{1.99}$$

where $E_3(x)$ *as in (1.74), and if additionally* $f \in L_p(\alpha, \beta)$, *then*

$$\rho_m(f) \le \int_{\alpha_1}^{\beta_1} G_3(x) \psi(x) \, dx, \tag{1.100}$$

where $G_3(x)$ *as in (1.75).*

We need

Remark 1.49 (Background). Let f be a convex function from $(0, +\infty)$ into $\mathbb{R}$ which is strictly convex at 1 with $f(1) = 0$. Let (X, A, λ) be a measure space, where λ is a finite or a σ-finite measure on (X, A). And let μ_1, μ_2 be two probability measures on (X, A) such that $\mu_1 \ll \lambda$, $\mu_2 \ll \lambda$ (absolutely continuous), e.g., $\lambda = \mu_1 + \mu_2$.

Denote by $p = \frac{d\mu_1}{d\lambda}$, $q = \frac{d\mu_2}{d\lambda}$ the (densities) Radon-Nikodym derivatives of μ_1, μ_2 with respect to λ. Here we suppose that

$$0 < \alpha \le \frac{p}{q} \le \beta, \text{ a.e. on } X \text{ and } \alpha \le 1 \le \beta.$$

The quantity

$$\Gamma_f(\mu_1, \mu_2) = \int_X q(x) f\left(\frac{p(x)}{q(x)} \right) d\lambda(x), \tag{1.101}$$

was introduced by I. Csiszar in 1967, see [9], and is called f-divergence of the probability measures μ_1 and μ_2. By Lemma 1.1 of [9], the integral (1.101) is well-defined and $\Gamma_f(\mu_1, \mu_2) \ge 0$ with equality only when $\mu_1 = \mu_2$. Furthermore $\Gamma_f(\mu_1, \mu_2)$ does not depend on the choice of λ. The concept of f-divergence was introduced first in [8] as a generalization of Kullback's "information for discrimination" or I-divergence (generalized entropy) [12, 13] and of Rényi's "information gain" (I-divergence of order δ) [15]. In fact the I-divergence of order 1 equals $\Gamma_{u \log_2 u} (\mu_1, \mu_2)$. The choice $f (x) = (u - 1)^2$ produces again a known measure of difference of distributions that is called χ^2-divergence. Of course the total variation distance $|\mu_1 - \mu_2| = \int_X |p (x) - q (x)| \, d\lambda (x)$ is equal to $\Gamma_{|u-1|} (\mu_1, \mu_2)$.

Here by supposing $f (1) = 0$ we can consider $\Gamma_f (\mu_1, \mu_2)$, the f-divergence as a measure of the difference between the probability measures μ_1, μ_2. The f-divergence is in general asymmetric in μ_1 and μ_2. But since f is convex and strictly convex at 1 so is

$$f^* (u) = u f \left(\frac{1}{u} \right) \tag{1.102}$$

and as in [9] we obtain

$$\Gamma_f (\mu_2, \mu_1) = \Gamma_{f^*} (\mu_1, \mu_2). \tag{1.103}$$

In information theory and statistics many other divergences are used which are special cases of the above general Csiszar f-divergence, e.g., the Hellinger distance D_H, α-distance D_α, Bhattacharyya distance D_B, Harmonic distance D_{Ha}, Jeffrey's distance D_J, triangular discrimination D_Δ, for all these see, e.g., [5, 10]. The problem of finding and estimating the proper distance (or difference or discrimination) of two probability distributions is one of the major ones in Probability Theory.

Here we provide a general probabilistic representation formula for $\Gamma_f (\mu_1, \mu_2)$. Then we present tight estimates for the remainder involving a variety of norms of the engaged functions. Also are implied some direct general approximations for the Csiszar's f-divergence. We give some applications.

We make

Remark 1.50. Here $0 < a < \alpha \leq \frac{p(x)}{q(x)} \leq \beta < b < +\infty$, a e. on X and $\alpha \leq 1 \leq \beta$. Also suppose that $f^{(l-1)}$ exists and is absolutely continuous on $[a, b]$, $l \in \mathbb{N}$. Furthermore f is convex from $(0, +\infty)$ into $\mathbb{R}$, strictly convex at 1 with $f (1) = 0$. Let $\omega \in L_1 (\mathbb{R})$, sup $p\omega \subset [\alpha, \beta]$, $\int_\mathbb{R} \omega (x) \, dx = 1$.

Then $\forall \, x \in (\alpha, \beta)$ we get by Theorem 1.10, as in (1.41), that

$$f (x) = \int_\alpha^\beta f (y) \omega (y) \, dy + Q^{l-1} f (x) + R_{0,l} f (x).$$

Therefore

$$f \left(\frac{p (x)}{q (x)} \right) = \int_\alpha^\beta f (y) \omega (y) \, dy + Q^{l-1} f \left(\frac{p (x)}{q (x)} \right) + R_{0,l} f \left(\frac{p (x)}{q (x)} \right),$$

a.e. on X.

Hence

$$q(x) f\left(\frac{p(x)}{q(x)}\right) = q(x) \int_\alpha^\beta f(y) \omega(y) \, dy$$
$$+ q(x) Q^{l-1} f\left(\frac{p(x)}{q(x)}\right) + q(x) R_{0,l} f\left(\frac{p(x)}{q(x)}\right),$$

a.e. on X.

Therefore we get the representation of f-divergence of μ_1 and μ_2,

$$\Gamma_f(\mu_1, \mu_2) = \int_X q(x) f\left(\frac{p(x)}{q(x)}\right) d\lambda(x) = \int_\alpha^\beta f(y) \omega(y) \, dy$$
$$+ \int_X q(x) Q^{l-1} f\left(\frac{p(x)}{q(x)}\right) d\lambda(x)$$
$$+ \int_X q(x) R_{0,l} f\left(\frac{p(x)}{q(x)}\right) d\lambda(x). \tag{1.104}$$

Call

$$Q_\Gamma := \int_X q(x) Q^{l-1} f\left(\frac{p(x)}{q(x)}\right) d\lambda(x), \tag{1.105}$$

and

$$R_\Gamma := \int_X q(x) R_{0,l} f\left(\frac{p(x)}{q(x)}\right) d\lambda(x). \tag{1.106}$$

We estimate Q_Γ and R_Γ.

If $\|\omega\|_{L_\infty(\mathbb{R})} < \infty$, we get by (1.29) that

$$|Q_\Gamma| \le \left(\sum_{k=1}^{l-1} \frac{(\beta - \alpha)^k}{k!} \left\|f^{(k)}\right\|_{L_1(\alpha,\beta)}\right) \|\omega\|_{L_\infty(\mathbb{R})}. \tag{1.107}$$

Notice if $l = 1$, then always $Q_\Gamma = 0$.

Next if again $\|\omega\|_{L_\infty(\mathbb{R})} < \infty$, then (by (1.30))

$$|Q_\Gamma| \le \left(\int_X q(x) \left(\sum_{k=1}^{l-1} \frac{\left(\beta - \frac{p(x)}{q(x)}\right)^{k+1} + \left(\frac{p(x)}{q(x)} - \alpha\right)^{k+1}}{(k+1)!}\right.\right.$$
$$\left.\left. \times \left\|f^{(k)}\right\|_{L_\infty(\alpha,\beta)}\right) d\lambda(x)\right) \|\omega\|_{L_\infty(\mathbb{R})}. \tag{1.108}$$

Let now $p, q > 1 : \frac{1}{p} + \frac{1}{q} = 1$ and again $\|\omega\|_{L_\infty(\mathbb{R})} < \infty$. Then (by (1.31))

$$|Q_\Gamma| \le \left(\int_X q(x) \left(\sum_{k=1}^{l-1} \left(\frac{\left(\beta - \frac{p(x)}{q(x)}\right)^{(kq+1)} + \left(\frac{p(x)}{q(x)} - \alpha\right)^{(kq+1)}}{kq+1} \right) \right.\right.$$

$$\left.\left. \times \frac{\|f^{(k)}\|_{L_p(\alpha,\beta)}}{k!} \right) d\lambda(x) \right)^{\frac{1}{q}} \|\omega\|_{L_\infty(\mathbb{R})} . \tag{1.109}$$

Next assume $\omega \in L_1(\mathbb{R})$, then (by (1.32))

$$|Q_\Gamma| \le \left(\sum_{k=1}^{l-1} \frac{\|f^{(k)}\|_{L_\infty(\alpha,\beta)} (\beta - \alpha)^k}{k!} \right) \|\omega\|_{L_1(\mathbb{R})} . \tag{1.110}$$

If $p, q > 1 : \frac{1}{p} + \frac{1}{q} = 1$ and $\omega \in L_q(\alpha, \beta)$, then (by (1.33))

$$|Q_\Gamma| \le \left(\sum_{k=1}^{l-1} \frac{(\beta - \alpha)^k}{k!} \|f^{(k)}\|_{L_p(\alpha,\beta)} \right) \|\omega\|_{L_q(\alpha,\beta)} . \tag{1.111}$$

Assume $p, q, r > 1 : \frac{1}{p} + \frac{1}{q} + \frac{1}{r} = 1$ and $\omega \in L_q(\alpha, \beta)$, then (by (1.34))

$$|Q_\Gamma| \le \left(\int_X q(x) \left(\sum_{k=1}^{l-1} \frac{\|f^{(k)}\|_{L_p(\alpha,\beta)}}{k!} \right.\right.$$

$$\left.\left. \times \left(\frac{\left(\beta - \frac{p(x)}{q(x)}\right)^{(kr+1)} + \left(\frac{p(x)}{q(x)} - \alpha\right)^{(kr+1)}}{kr+1} \right)^{\frac{1}{r}} \right) d\lambda(x) \right)$$

$$\times \|\omega\|_{L_q(\alpha,\beta)} . \tag{1.112}$$

We make

Remark 1.51 (continuation of Remark 1.50). Here $l > 1$, $\omega \in C^{(l-2)}(\mathbb{R})$, sup $p\omega \subset [\alpha, \beta]$, $\int_\mathbb{R} \omega(x) dx = 1$, and $\omega^{(l-2)}$ is absolutely continuous on $[a, b]$. Then (by (1.35))

$$Q_\Gamma = \int_X q(x) \left(\sum_{k=1}^{l-1} \frac{(-1)^k}{k!} \int_\alpha^\beta \left(\left[\left(\frac{p(x)}{q(x)} - y\right)^k \omega(y) \right]_y^{(k)} \right) f(y) dy \right) d\lambda(x) . \tag{1.113}$$

Hence by (1.37) we obtain

$|Q_\Gamma| \leq$ min of

$$
\begin{cases}
\left(\int_X q(x) \left(\sum_{k=1}^{l-1} \frac{1}{k!} \left\| \left[\left(\frac{p(x)}{q(x)} - y \right)^k \omega(y) \right]_y^{(k)} \right\|_\infty \right) d\lambda(x) \right) \|f\|_{L_1(\alpha,\beta)}, \\[4mm]
\left(\int_X q(x) \left(\sum_{k=1}^{l-1} \frac{1}{k!} \left\| \left[\left(\frac{p(x)}{q(x)} - y \right)^k \omega(y) \right]_y^{(k)} \right\|_{L_1(\alpha,\beta)} \right) d\lambda(x) \right) \|f\|_{L_\infty(\alpha,\beta)}, \\[4mm]
\text{when } p, q > 1 : \frac{1}{p} + \frac{1}{q} = 1, \text{ we have} \\[2mm]
\left(\int_X q(x) \left(\sum_{k=1}^{l-1} \frac{1}{k!} \left\| \left[\left(\frac{p(x)}{q(x)} - y \right)^k \omega(y) \right]_y^{(k)} \right\|_{L_q(\alpha,\beta)} \right) d\lambda(x) \right) \|f\|_{L_p(\alpha,\beta)}.
\end{cases}
$$

$$(1.114)$$

We also make

Remark 1.52   (another continuation of Remark 1.50). Here we estimate the remainder R_Γ of (1.104). By (1.23) and (1.106), we obtain

$$
|R_\Gamma| \leq \frac{\|\omega\|_{L_1(a,b)} \left\| f^{(l)} \right\|_{L_1(\alpha,\beta)} (\beta - \alpha)^{l-1}}{(l-1)!}.
\tag{1.115}
$$

If $f^{(l)} \in L_\infty(\alpha, \beta)$, then (by (1.24)) we obtain

$$
|R_\Gamma| \leq \frac{\|\omega\|_{L_1(a,b)} \left\| f^{(l)} \right\|_{L_\infty(\alpha,\beta)}}{l!} \left(\int_X q(x) \left(\left(\beta - \frac{p(x)}{q(x)} \right)^l \right.\right.
$$

$$
\left.\left. + \left(\frac{p(x)}{q(x)} - \alpha \right)^l \right) d\lambda(x) \right).
\tag{1.116}
$$

Let now $p, q > 1 : \frac{1}{p} + \frac{1}{q} = 1$. Here $f^{(l)} \in L_p(\alpha, \beta)$, then (by (1.25)) we get

$$
|R_\Gamma| \leq \frac{\|\omega\|_{L_1(a,b)} \left\| f^{(l)} \right\|_{L_p(\alpha,\beta)}}{(q(l-1)+1)^{\frac{1}{q}} (l-1)!} \left(\int_X q(x) \left(\left(\beta - \frac{p(x)}{q(x)} \right)^{(q(l-1)+1)} \right.\right.
$$

$$
\left.\left. + \left(\frac{p(x)}{q(x)} - \alpha \right)^{(q(l-1)+1)} \right)^{\frac{1}{q}} d\lambda(x) \right).
\tag{1.117}
$$

Finally we observe that

$$
\Gamma_f(\mu_1, \mu_2) - \int_\alpha^\beta f(y) \omega(y) \, dy = Q_\Gamma + R_\Gamma,
\tag{1.118}
$$

and

$$T := \left| \Gamma_f \left(\mu_1, \mu_2 \right) - \int_\alpha^\beta f \left(y \right) \omega \left(y \right) dy \right| \leq |Q_\Gamma| + |R_\Gamma|. \qquad (1.119)$$

Then one by the above estimates of $|Q_\Gamma|$ and $|R_\Gamma|$ can estimate T, in a number of cases.

1.4 Applications

Example 1.53. Let $V := \{x \in \mathbb{R} : |x - x_0| < \rho\}$, $x_0 \in \mathbb{R}$, and

$$\varphi (x) := \begin{cases} e^{-\left(1 - \frac{(x-x_0)^2}{\rho^2}\right)^{-1}}, & \text{if } |x - x_0| < \rho, \\ 0, & \text{if } |x - x_0| \geq \rho. \end{cases} \qquad (1.120)$$

Call $c := \int_\mathbb{R} \varphi (x) \, dx > 0$, then $\Phi (x) := \frac{1}{c} \varphi (x) \in C_0^\infty (\mathbb{R})$ (space of continuously infinitely many times differentiable functions of compact support) with $\sup p\Phi = \overline{V}$ and $\int_{-\infty}^\infty \Phi (x) \, dx = 1$ and $\max |\Phi| \leq cons \tan t \cdot \rho^{-1}$. We call Φ a cut-off function.

One for this chapter's results by choosing $\omega (x) = \Phi (x)$ or $\omega (x) = \frac{1}{2\rho}$, etc., can give lots of applications. Due to lack of space we avoid it.

Instead, selectively, we give some special cases inequalities. We start with Chebyshev–Grüss-type inequalities.

Corollary 1.54 (to Theorem 1.22). *Let* $f, g \in C^1 ([a, b])$, $[a, b] \subset \mathbb{R}$, $(\alpha, \beta) \subset (a, b)$. *Let also* $\omega \in L_\infty (\mathbb{R})$, $\sup p\omega \subset [\alpha, \beta]$, $\int_\mathbb{R} \omega (x) \, dx = 1$. *Then*

$$\left| \int_\alpha^\beta \omega (x) f (x) g (x) \, dx - \left(\int_\alpha^\beta \omega (x) f (x) \, dx \right) \left(\int_\alpha^\beta \omega (x) g (x) \, dx \right) \right|$$

$$\leq \|\omega\|_{L_1(\mathbb{R})} \|\omega\|_{\infty,(\alpha,\beta)} \frac{(\beta - \alpha)^2}{2}$$

$$\times \left(\|g\|_{\infty,(\alpha,\beta)} \|f'\|_{\infty,(\alpha,\beta)} + \|f\|_{\infty,(\alpha,\beta)} \|g'\|_{\infty,(\alpha,\beta)} \right). \qquad (1.121)$$

If $f = g$, *then*

$$\left| \int_\alpha^\beta \omega (x) f^2 (x) \, dx - \left(\int_\alpha^\beta \omega (x) f (x) \, dx \right)^2 \right|$$

$$\leq \|\omega\|_{L_1(\mathbb{R})} \|\omega\|_{\infty,(\alpha,\beta)} (\beta - \alpha)^2 \|f\|_{\infty,(\alpha,\beta)} \|f'\|_{\infty,(\alpha,\beta)}. \qquad (1.122)$$

Corollary 1.55 (to Theorem 1.23). *Let* $f \in \left(W_1^1\right)^{loc}(a,b)$; $a,b \in \mathbb{R}$; $(\alpha, \beta) \subset (a,b)$, $\omega(x) := \frac{1}{\beta - \alpha}$ *for* $x \in [\alpha, \beta]$, *and zero elsewhere. Then*

$$\left| \frac{1}{\beta - \alpha} \int_\alpha^\beta f^2(x)\,dx - \frac{1}{(\beta - \alpha)^2} \left(\int_\alpha^\beta f(x)\,dx \right)^2 \right| \leq \frac{\|f\|_{L_1(\alpha,\beta)} \left\| f_w^{(1)} \right\|_{L_1(\alpha,\beta)}}{(\beta - \alpha)}.$$

(1.123)

We continue with an Ostrowski-type inequality.

Corollary 1.56 (to Theorem 1.30). *All as in Theorem 1.30. Case of* $l = 1$. *Then, for any* $x \in (\alpha, \beta)$ *(or for almost every* $x \in (\alpha, \beta)$, *respectively), we get*

$$\left| f(x) - \int_\alpha^\beta f(y)\,\omega(y)\,dy \right| \leq \|\omega\|_{L_1(\mathbb{R})} \left\| \overrightarrow{f'} \right\|_{L_1(\alpha,\beta)}.$$

(1.124)

Next comes a comparison of means inequality.

Corollary 1.57. *All here as in Corollary 1.56 and Remark 1.38. Then*

$$\left| \int_{\alpha_1}^{\beta_1} f(x)\,\psi(x)\,dx - \int_\alpha^\beta f(y)\,\omega(y)\,dy \right| \leq \|\omega\|_{L_1(\mathbb{R})} \left\| \overrightarrow{f'} \right\|_{L_1(\alpha,\beta)}.$$

(1.125)

Proof. By (1.124). ∎

We finish with an application of f-divergence.

Remark 1.58. All here as in Background 1.49 and Remark 1.50. Case of $l = 1$. By (1.104) we get

$$\Gamma_f(\mu_1, \mu_2) = \int_\alpha^\beta f(y)\,\omega(y)\,dy + \int_X q(x)\,R_{0,1} f\left(\frac{p(x)}{q(x)} \right) d\lambda(x).$$

(1.126)

That is here

$$R_\Gamma = \int_X q(x)\,R_{0,1} f\left(\frac{p(x)}{q(x)} \right) d\lambda(x).$$

(1.127)

By (1.115) here we get that

$$|R_\Gamma| \leq \|\omega\|_{L_1(a,b)} \|f'\|_{L_1(\alpha,\beta)}.$$

(1.128)

If $f' \in L_\infty(\alpha, \beta)$, then here we get

$$|R_\Gamma| \leq \|\omega\|_{L_1(a,b)} \|f'\|_{L_\infty(\alpha,\beta)} (\beta - \alpha).$$

(1.129)

Let now $p, q > 1 : \frac{1}{p} + \frac{1}{q} = 1$ and assume $f' \in L_p(\alpha, \beta)$, then here we obtain

$$|R_\Gamma| \leq \|\omega\|_{L_1(a,b)} \|f'\|_{L_p(\alpha,\beta)} (\beta - \alpha)^{\frac{1}{q}}. \tag{1.130}$$

Also notice here that

$$K := \Gamma_f(\mu_1, \mu_2) - \int_\alpha^\beta f(y)\,\omega(y)\,\mathrm{d}y = R_\Gamma, \tag{1.131}$$

($l = 1$ case).
So the estimates (1.128)–(1.130) are also estimates for K.

References

1. G.A. Anastassiou, *Quantitative Approximations*, Chapman & Hall/CRC, Boca Raton, New York, 2000.
2. G.A. Anastassiou, *Probabilistic Inequalities*, World Scientific, Singapore, New Jersey, 2010.
3. G.A. Anastassiou, *Advanced Inequalities*, World Scientific, Singapore, New Jersey, 2011.
4. G.A. Anastassiou, *Univariate Inequalities based on Sobolev Representations*, Studia Mathematica-Babes Bolyai, accepted 2011.
5. N.S. Barnett, P. Cerone, S.S. Dragomir, A. Sofo, *Approximating Csiszar's f-divergence by the use of Taylor's formula with integral remainder*, (paper #10, pp. 16), Inequalities for Csiszar's f-Divergence in Information Theory, S.S. Dragomir (ed.), Victoria University, Melbourne, Australia, 2000. On line: http://rgmia.vu.edu.au
6. V. Burenkov, *Sobolev spaces and domains*, B.G. Teubner, Stuttgart, Leipzig, 1998.
7. P.L. Chebyshev, *Sur les expressions approximatives des integrales definies par les autres prises entre les mêmes limites*, Proc. Math. Soc. Charkov, 2(1882), 93-98.
8. I. Csiszar, *Eine Informationstheoretische Ungleichung und ihre Anwendung auf den Beweis der Ergodizität von Markoffschen Ketten*, Magyar Tud. Akad. Mat. Kutato Int. Közl. 8 (1963), 85-108.
9. I. Csiszar, *Information-type measures of difference of probability distributions and indirect observations*, Studia Math. Hungarica 2 (1967), 299-318.
10. S.S. Dragomir (ed.), *Inequalities for Csiszar f-Divergence in Information Theory*, Victoria University, Melbourne, Australia, 2000. On-line: http://rgmia.vu.edu.au
11. G. Grüss, *Über das Maximum des absoluten Betrages von* $\left[\left(\frac{1}{b-a}\right) \int_a^b f(x)\,g(x)\,\mathrm{d}x - \left(\frac{1}{(b-a)^2} \int_a^b f(x)\,\mathrm{d}x \int_a^b g(x)\,\mathrm{d}x\right)\right]$, Math. Z. 39 (1935), pp. 215-226.
12. S. Kullback, *Information Theory and Statistics*, Wiley, New York, 1959.
13. S. Kullback, R. Leibler, *On information and sufficiency*, Ann. Math. Statist., 22 (1951), 79-86.
14. A. Ostrowski, *Uber die Absolutabweichung einer differentiabaren Funcktion von ihrem Integralmittelwert*, Comment. Math. Helv. 10(1938), 226-227.
15. A. Rényi, *On measures of entropy and information*, Proceedings of the 4th Berkeley Symposium on Mathematical Statistics and Probability, I, Berkeley, CA, 1960, 547-561.

Chapter 2
Multivariate Integral Inequalities Deriving from Sobolev Representations

Here we present very general multivariate tight integral inequalities of Chebyshev–Grüss, Ostrowski types and of comparison of integral means. These rely on the well-known Sobolev integral representation of a function. The inequalities engage ordinary and weak partial derivatives of the involved functions. We give also applications. On the way to prove the main results we obtain important estimates for the averaged Taylor polynomials and remainders of Sobolev integral representations. The exposed results are thoroughly discussed. This chapter relies on [4].

2.1 Introduction

This chapter is greatly motivated by the following theorems:

Theorem A (Chebychev, 1882, [7]). *Let* $f, g : [a, b] \rightarrow R$ *absolutely continuous functions. If* $f', g' \in L_\infty ([a, b])$, *then*

$$\left| \frac{1}{b-a} \int_a^b f(x) g(x) \, dx - \frac{1}{(b-a)^2} \left(\int_a^b f(x) \, dx \right) \left(\int_a^b g(x) \, dx \right) \right|$$

$$\leq \frac{1}{12} (b-a)^2 \, \|f'\|_\infty \, \|g'\|_\infty \, .$$

Theorem B (G. Grüss, 1935, [8]). *Let* f, g *integrable functions from* $[a, b] \rightarrow R$, *such that* $m \leq f(x) \leq M$, $\rho \leq g(x) \leq \sigma$, *for all* $x \in [a, b]$, *where* $m, M, \rho, \sigma \in \mathbb{R}$. *Then*

$$\left| \frac{1}{b-a} \int_a^b f(x) g(x) \, dx - \frac{1}{(b-a)^2} \left(\int_a^b f(x) \, dx \right) \left(\int_a^b g(x) \, dx \right) \right|$$

$$\leq \frac{1}{4} (M-m)(\sigma - \rho) \, .$$

In 1938, A. Ostrowski [9] proved.

G.A. Anastassiou, *Inequalities Based on Sobolev Representations*,
SpringerBriefs in Mathematics 2, DOI 10.1007/978-1-4614-0201-5_2,
© George A. Anastassiou

Theorem C. *Let* $f : [a, b] \rightarrow R$ *be continuous on* $[a, b]$ *and differentiable on* (a, b) *whose derivative* $f' : (a, b) \rightarrow R$ *is bounded on* (a, b), *i.e.,* $\|f'\|_\infty = \sup_{t \in (a,b)} |f'(t)| < +\infty$. *Then*

$$\left| \frac{1}{b-a} \int_a^b f(t)\, dt - f(x) \right| \leq \left[\frac{1}{4} + \frac{\left(x - \frac{a+b}{2}\right)^2}{(b-a)^2} \right] \cdot (b-a) \|f'\|_\infty,$$

for any $x \in [a, b]$. *The constant* $\frac{1}{4}$ *is the best possible.*

See also [1–3] for related works that inspired as well this chapter.

In this chapter using the Sobolev-type representation formulae, see Theorems 2.6, 2.8, 2.11 and 2.23, also Corollaries 2.12 and 2.13, we estimate first their remainders and then the involved averaged Taylor polynomials.

Based on these estimates we establish lots of very tight inequalities on $\mathbb{R}^n$, $n \in \mathbb{N}$, of Chebyshev–Grüss type, Ostrowski type and of Comparison of integral means with applications. The results involve ordinary and weak partial derivatives and they go to all possible directions using various norms. All of our machinery comes from the excellent monograph by V. Burenkov, [6].

2.2   Background

Here we follow [6].

For a measurable nonempty set $\Omega \subset \mathbb{R}^n$, $n \in \mathbb{N}$ we shall denote by $L_p^{loc}(\Omega)$ $(1 \leq p \leq \infty)$ - the set of functions defined on Ω such that for each compact $K \subset \Omega$, $f \in L_p(K)$.

Definition 2.1. Let $\Omega \subset \mathbb{R}^n$ be an open set, $\alpha \in \mathbb{Z}_+^n$, $\alpha \neq 0$ and $f, g \in L_1^{loc}(\Omega)$. The function g is a weak derivative of the function f of order α on Ω (briefly $g = D_w^\alpha f$) if $\forall \varphi \in C_0^\infty(\Omega)$ (i.e., $\varphi \in C^\infty(\Omega)$ compactly supported in Ω)

$$\int_\Omega f D^\alpha \varphi dx = (-1)^{|\alpha|} \int_\Omega g\varphi\, dx. \tag{2.1}$$

Definition 2.2. $W_p^l(\Omega)$ $(l \in \mathbb{N}, 1 \leq p \leq \infty)$ – Sobolev space, which is the Banach space of functions $f \in L_p(\Omega)$ such that $\forall \alpha \in \mathbb{Z}_+^n$ where $|\alpha| \leq l$ the weak derivatives $D_w^\alpha f$ exist on Ω and $D_w^\alpha f \in L_p(\Omega)$, with the norm

$$\|f\|_{W_p^l(\Omega)} = \sum_{|\alpha| \leq l} \|D_w^\alpha f\|_{L_p(\Omega)}. \tag{2.2}$$

Definition 2.3. For $l \in \mathbb{N}$, we define the Sobolev-type local space $\left(W_1^l\right)^{(loc)}(\Omega) :=$ $\{f : \Omega \rightarrow \mathbb{R} : f \in L_{loc}^1(\Omega)$ and all f-distributional partials of orders $\leq l$ belong to $L_{loc}^1(\Omega)\} = \{f \in L_1^{loc}(\Omega) :$ for each open set G compactly embedded into Ω, $f \in W_1^l(G)\}$.

Definition 2.4. A domain $\Omega \subset \mathbb{R}^n$ is called star-shaped with respect to the point $y \in \Omega$ if $\forall\, x \in \Omega$ the closed interval line segment $[x, y] \subset \Omega$. A domain $\Omega \subset \mathbb{R}^n$ is called star-shaped with respect to an open ball $B \subset \Omega$ if $\forall\, y \in B$ and $\forall\, x \in \Omega$ we have $[x, y] \subset \Omega$.

We call the set

$$V_x = V_{x,B} = \cup_{y \in B} (x, y) = \text{ convex hull of } \{x\} \cup B,$$

a conic body with vertex x constructed on the open ball B (if $x \in B$, then $V_x = B$; if $\overline{B} \subset \Omega$ and $x \in \overline{B}$, then $V_x = B$).

In fact V_x for otherwise is the region consisting of B and the part of the cone with vertex at x, tangent to the sphere of B, which lies between x and B.

Next comes the multidimensional Taylor's formula.

Theorem 2.5. *Let $\Omega \subset \mathbb{R}^n$ be a domain star-shaped with respect to the point $x_0 \in \Omega$, $l \in \mathbb{N}$ and $f \in C^l(\Omega)$. Then $\forall\, x \in \Omega$*

$$f(x) = \sum_{|\alpha| < l} \frac{(D^\alpha f)(x_0)}{\alpha!} (x - x_0)^\alpha + l \sum_{|\alpha| = l}$$

$$\times \frac{(x - x_0)^\alpha}{\alpha!} \int_0^1 (1 - t)^{l-1} (D^\alpha f)(x_0 + t(x - x_0))\, dt \qquad (2.3)$$

(here we mean $x_0 + t(x - x_0) = (x_{01} + t(x_1 - x_{01}), \dots, x_{0n} + t(x_n - x_{0n}))$, $\alpha = (\alpha_1, \dots, \alpha_n) \in \mathbb{Z}_+^n$, $|\alpha| = \sum_{i=1}^n \alpha_i$, $\alpha! = \alpha_1! \dots \alpha_n!$, $(x - x_0)^\alpha = (x_1 - x_{01})^{\alpha_1} \dots (x_n - x_{0n})^{\alpha_n})$. Here $|\cdot|$ stands for the Euclidean norm: $|x| = \sqrt{\sum_{i=1}^n x_i^2}$, $x := (x_1, \dots, x_n)$.

Next we mention the Sobolev representations.

Theorem 2.6. *Let $\Omega \subset \mathbb{R}^n$ be a domain star-shaped with respect to the open ball $B = B(x_0, r)$ such that $\overline{B} \subset \Omega$, $\omega \in L_1(\mathbb{R}^n)$, the support supp $\omega \subset \overline{B}$, $\int_{\mathbb{R}^n} \omega(x)\, dx = 1$, $l \in \mathbb{N}$ and $f \in C^l(\Omega)$. Then for every $x \in \Omega$*

$$f(x) = \sum_{|\alpha| < l} \frac{1}{\alpha!} \int_B (D^\alpha f)(y)(x - y)^\alpha \omega(y)\, dy + l \sum_{|\alpha| = l} \frac{1}{\alpha!} \int_B (x - y)^\alpha \omega(y)$$

$$\times \left(\int_0^1 (1 - t)^{l-1} (D^\alpha f)(y + t(x - y))\, dt \right) dy. \qquad (2.4)$$

Proof. We write (2.3) for $x, x_0 = y$, multiply it both sides by $\omega(y)$ and integrate on B with respect to y. ∎

Call $\|D^\alpha f\|_{\infty,l,B}^{\max} := \max_{|\alpha|=l} \{\|D^\alpha f\|_{\infty,B}\}$, where $\|\cdot\|_{\infty,B}$ is the supremum norm on B, $d :=$ diameter of B.

Proposition 2.7. *Same assumption as in Theorem 2.6, $x \in B$. Then*

$$R_1 := |Remainder\,(4)| \leq \frac{(nd)^l \, \|\omega\|_{L_1(\mathbb{R}^n)} \, \|D^\alpha f\|_{\infty,l,B}^{\max}}{l!}. \tag{2.5}$$

Proof. We have that

$$\left| l \sum_{|\alpha|=l} \frac{1}{\alpha!} \int_B (x-y)^\alpha \, \omega\,(y) \left(\int_0^1 (1-t)^{l-1} \, (D^\alpha f)\,(y+t\,(x-y)) \, dt \right) dy \right|$$

$$\leq l \sum_{|\alpha|=l} \frac{1}{\alpha!} \int_B |(x-y)^\alpha| \cdot |\omega\,(y)| \left(\int_0^1 (1-t)^{l-1} \, |(D^\alpha f)\,(y+t\,(x-y))| \, dt \right) dy$$

$$\leq \|D^\alpha f\|_{\infty,l,B}^{\max} \left(\sum_{|\alpha|=l} \frac{1}{\alpha!} \int_B |(x-y)^\alpha| \cdot |\omega\,(y)| \, dy \right)$$

$$\leq \|D^\alpha f\|_{\infty,l,B}^{\max} \cdot d^l \cdot \left(\sum_{|\alpha|=l} \frac{1}{\alpha!} \int_B |\omega\,(y)| \, dy \right)$$

$$= \frac{\|D^\alpha f\|_{\infty,l,B}^{\max} \cdot d^l \cdot \|\omega\|_{L_1(B)}}{l!} \left(\sum_{|\alpha|=l} \frac{l!}{\alpha!} \right)$$

$$= \frac{\|D^\alpha f\|_{\infty,l,B}^{\max} \cdot (d \cdot n)^l \, \|\omega\|_{L_1(B)}}{l!}.$$

     ∎

From [6], p. 104, we mention

Theorem 2.8. *Let $\Omega \subset \mathbb{R}^n$ be a domain star-shaped with respect to the open ball $B = B\,(x_0, r)$ such that $\overline{B} \subset \Omega$,*

$$\omega \in L_1\,(\mathbb{R}^n), \;\; supp\,\omega \subset \overline{B}, \; \int_{\mathbb{R}^n} \omega\,(x) \, dx = 1, \tag{2.6}$$

$l \in \mathbb{N}$ and $f \in C^l\,(\Omega)$. Then for every $x \in \Omega$

$$f\,(x) = \sum_{|\alpha|<l} \frac{1}{\alpha!} \int_B (D^\alpha f)\,(y)\,(x-y)^\alpha \, \omega\,(y)\, dy$$

$$+ \sum_{|\alpha|=l} \int_{V_x} \frac{(D^\alpha f)\,(y)}{|x-y|^{n-l}} w_\alpha\,(x,y)\, dy, \tag{2.7}$$

where for $x, y \in \mathbb{R}^n$, $x \neq y$,

$$w_\alpha (x, y) := \frac{|\alpha|}{\alpha!} \frac{(x-y)^\alpha}{|x-y|^{|\alpha|}} w(x, y), \tag{2.8}$$

and

$$w(x, y) := \int_{|x-y|}^\infty \omega\left(x + \rho \frac{y-x}{|y-x|}\right) \rho^{n-1} d\rho, \tag{2.9}$$

(for $x = y \in \Omega$ we define $w_\alpha(x, x) = w(x, x) = 0$).

Remark 2.9. By (2.4) and (2.7) we derive

$$l \sum_{|\alpha|=l} \frac{1}{\alpha!} \int_B (x-y)^\alpha \, \omega(y) \left(\int_0^1 (1-t)^{l-1} (D^\alpha f)(y + t(x-y)) \, dt\right) dy$$

$$= \sum_{|\alpha|=l} \int_{V_x} \frac{(D^\alpha f)(y)}{|x-y|^{n-l}} w_\alpha(x, y) \, dy. \tag{2.10}$$

By Proposition 2.7 we obtain

$$\left|\sum_{|\alpha|=l} \int_{V_x} \frac{(D^\alpha f)(y)}{|x-y|^{n-l}} w_\alpha(x, y) \, dy\right| \leq \frac{(nd)^l \, \|\omega\|_{L_1(\mathbb{R}^n)} \, \|D^\alpha f\|_{\infty,l}^{\max}}{l!}. \tag{2.11}$$

Remark 2.10. Let $D =$ diameter of Ω be finite, i.e., Ω is bounded. By [6] we have

$$\|w(x, y)\|_{C(\mathbb{R}^n \times \mathbb{R}^n)} \leq \|\omega\|_{L_\infty(\mathbb{R}^n)} D^{n-1} d, \tag{2.12}$$

and $\forall \, \alpha \in \mathbb{Z}_+^n$ satisfying $|\alpha| = l$

$$\|w_\alpha(x, y)\|_{C(\mathbb{R}^n \times \mathbb{R}^n)} \leq \|\omega\|_{L_\infty(\mathbb{R}^n)} n D^{n-1} d. \tag{2.13}$$

Notice $\|w(x, y)\|_{C(\mathbb{R}^n \times \mathbb{R}^n)} \leq \|\omega\|_{L_\infty(\mathbb{R}^n)} d^n$ and $\|w_\alpha(x, y)\|_{C(\mathbb{R}^n \times \mathbb{R}^n)} \leq \|\omega\|_{L_\infty(\mathbb{R}^n)} n d^n$, if $\Omega = B$. Hence, if ω is bounded, then for bounded Ω the functions w and w_α are bounded on $\mathbb{R}^n \times \mathbb{R}^n$. Also by [6], if Ω is unbounded, then w, w_α are bounded on $K \times \mathbb{R}^n$ for each compact K.

If $\omega \in C^\infty(\mathbb{R}^n)$, then $w(x, y)$, $w_\alpha(x, y)$ have continuous derivatives of all orders $\forall \, x, y \in \mathbb{R}^n : x \neq y$ and at the points (x, x), where $x \notin \overline{B}$ they are discontinuous, see [6].

Finally, we give the very general Sobolev representation, see [6].

Theorem 2.11. *Let $\Omega \subset \mathbb{R}^n$ be a domain star-shaped with respect to the open ball $B = B(x_0, r)$ such that $\overline{B} \subset \Omega$,*

$$\omega \in L_\infty(\mathbb{R}^n), \quad supp\, \omega \subset \overline{B}, \quad \int_{\mathbb{R}^n} \omega(x)\,dx = 1, \tag{2.14}$$

$l \in \mathbb{N}$ and $f \in \left(W_1^l\right)^{loc}(\Omega)$. Then for almost every $x \in \Omega$

$$f(x) = \sum_{|\alpha|<l} \frac{1}{\alpha!} \int_B \left(D_w^\alpha f\right)(y)(x-y)^\alpha \omega(y)\,dy$$

$$+ \sum_{|\alpha|=l} \int_{V_x} \frac{\left(D_w^\alpha f\right)(y)}{|x-y|^{n-l}} w_\alpha(x,y)\,dy. \tag{2.15}$$

Corollary 2.12 ([6]). *Let $\Omega \subset \mathbb{R}^n$ be a domain star-shaped with respect to the open ball $B = B(x_0, r)$ such that $\overline{B} \subset \Omega$,*

$$\omega \in C_0^\infty(\Omega), \quad supp\, \omega \subset \overline{B}, \quad \int_{\mathbb{R}^n} \omega(x)\,dx = 1. \tag{2.16}$$

Then $\forall\, f \in C^l(\Omega)$ for every $x \in \Omega$ and $\forall\, f \in \left(W_1^l\right)^{loc}(\Omega)$ for almost every $x \in \Omega$

$$f(x) = \int_B \left(\sum_{|\alpha|<l} \frac{(-1)^{|\alpha|}}{\alpha!} D_y^\alpha \left[(x-y)^\alpha \omega(y)\right] \right) f(y)\,dy$$

$$+ \sum_{|\alpha|=l} \int_{V_x} \frac{\left(D^\alpha f\right)(y)}{|x-y|^{n-l}} w_\alpha(x,y)\,dy \tag{2.17}$$

with $D_w^\alpha f$ replacing $D^\alpha f$ in the case of $f \in \left(W_1^l\right)^{loc}(\Omega)$.

Next $\alpha \geq \beta$ means $\alpha_i \geq \beta_i, i = 1, ..., n$ and $\alpha - \beta \in \mathbb{Z}_+$.

Corollary 2.13 ([6]). *Under the assumptions of Corollary 2.12, let $\beta \in \mathbb{Z}_+^n$ and $0 < |\beta| < l$. Then $\forall\, f \in C^l(\Omega)$ for every $x \in \Omega$ and $\forall\, f \in \left(W_1^l\right)^{loc}(\Omega)$ for almost every $x \in \Omega$*

$$\left(D^\beta f\right)(x) = \int_B \left(\sum_{|\alpha|<l-|\beta|} \frac{(-1)^{|\alpha|+|\beta|}}{\alpha!} D_y^{\alpha+\beta} \left[(x-y)^\alpha \omega(y)\right] \right) f(y)\,dy$$

$$+ \sum_{|\alpha|=l, \alpha \geq \beta} \int_{V_x} \frac{\left(D^\alpha f\right)(y)}{|x-y|^{n-l+|\beta|}} w_{\alpha-\beta}(x,y)\,dy \tag{2.18}$$

with $D_w^\beta f$ replacing $D^\beta f$ and $D_w^\alpha f$ replacing $D^\alpha f$ if $f \in \left(W_1^l\right)^{loc}(\Omega)$.

Remark 2.14. Again $d = diamB$, $D = diam\Omega$. We suppose $\|\omega\|_{L_\infty(\mathbb{R}^n)} < \infty$. Here $\overline{D}^\alpha f$ could mean either $D^\alpha f$ or $D_w^\alpha f$. Then

$$R_2 := \left| \sum_{|\alpha|=l} \int_{V_x} \frac{\left(\overline{D}^\alpha f\right)(y)}{|x-y|^{n-l}} w_\alpha(x,y)\,dy \right|$$

$$\leq \sum_{|\alpha|=l} \int_{V_x} |x-y|^{l-n} \left|\left(\overline{D}^\alpha f\right)(y)\right| |w_a(x,y)|\,dy$$

$$\leq \left(\sum_{|\alpha|=l} \int_{V_x} |x-y|^{l-n} \left|\left(\overline{D}^\alpha f\right)(y)\right|\,dy \right) \|\omega\|_{L_\infty(\mathbb{R}^n)}\, n\, d\, D^{n-1}, \qquad (2.19)$$

if $D < \infty$.

Next we assume $x \in B$, $l \geq n$, then we retake $\Omega = B$, i.e., $d = D$, etc., and thus

$$R_2 \leq \sum_{|\alpha|=l} \left(\int_B \left|\left(\overline{D}^\alpha f\right)(y)\right|\,dy \right) \|\omega\|_{L_\infty(\mathbb{R}^n)}\, n\, d^l$$

$$= \left(\sum_{|\alpha|=l} \left\|\overline{D}^\alpha f\right\|_{L_1(B)} \right) \|\omega\|_{L_\infty(\mathbb{R}^n)}\, n\, d^l.$$

That is, we proved for $l \geq n$, $x \in B$, that

$$\left| \sum_{|\alpha|=l} \int_B \frac{\left(\overline{D}^\alpha f\right)(y)}{|x-y|^{n-l}} w_\alpha(x,y)\,dy \right| \leq \left(\sum_{|\alpha|=l} \left\|\overline{D}^\alpha f\right\|_{L_1(B)} \right) \|\omega\|_{L_\infty(\mathbb{R}^n)}\, n\, d^l.$$

$$(2.20)$$

Again we assume $x \in B$, $l \geq n$ and $\left\|\overline{D}^\alpha f\right\|_{L_\infty(B)} < \infty$ for all $\alpha : |\alpha| = l$ (which is true for $D^\alpha f$ always by $f \in C^l(\Omega)$). Then

$$R_2 \leq \left(\sum_{|\alpha|=l} \left\|\overline{D}^\alpha f\right\|_{L_\infty(B)} \right) Vol(B) \|\omega\|_{L_\infty(\mathbb{R}^n)}\, n\, d^l =: (*).$$

We know that

$$Vol(B) = \frac{\pi^{\frac{n}{2}}}{\Gamma\left(\frac{n}{2}+1\right)} r^n = \frac{\pi^{\frac{n}{2}}}{\Gamma\left(\frac{n}{2}+1\right)} \frac{d^n}{2^n},$$

where Γ is the gamma function.

Therefore

$$
(*) = \left(\sum_{|\alpha|=l} \left\| \overline{D}^{\alpha} f \right\|_{L_{\infty}(B)} \right) \| \omega \|_{L_{\infty}(\mathbb{R}^n)} \frac{n \pi^{\frac{n}{2}}}{2^n \Gamma \left(\frac{n}{2} + 1 \right)} d^{l+n}.
$$

So we have proved if $l \geq n$, $x \in B$, and $\left\| \overline{D}^{\alpha} f \right\|_{L_{\infty}(B)} < \infty$ for all $\alpha : |\alpha| = l$ that

$$
\left| \sum_{|\alpha|=l} \int_B \frac{\left(\overline{D}^{\alpha} f \right)(y)}{|x - y|^{n-l}} w_{\alpha}(x, y) \, dy \right|
$$

$$
\leq \left(\sum_{|\alpha|=l} \left\| \overline{D}^{\alpha} f \right\|_{L_{\infty}(B)} \right) \| \omega \|_{L_{\infty}(\mathbb{R}^n)} \frac{n \pi^{\frac{n}{2}}}{2^n \Gamma \left(\frac{n}{2} + 1 \right)} d^{l+n}. \qquad (2.21)
$$

We use Lemma (4.3.1), p. 100 of [5].
It follows

Lemma 2.15. *If $f \in L_p(\Omega^*)$, $1 < p < \infty$, Ω^* is a region of diameter $d_* > 0$, and $m > \frac{n}{p}$, then*

$$
\int_{\Omega} |x - z|^{m-n} |f(z)| \, dz \leq c_p d_*^{m - \frac{n}{p}} \| f \|_{L_p(\Omega^*)}, \qquad (2.22)
$$

$\forall x \in \Omega^*$, *where c_p is a constant depending only on p.*

We make

Remark 2.16 (continuing from Remark 2.14). We assume now that $\overline{D}^{\alpha} f \in L_p(B)$, $|\alpha| = l$, $1 < p < \infty$, $l > \frac{n}{p}$. Then by (2.22),

$$
\int_B |x - y|^{l-n} \left| \left(\overline{D}^{\alpha} f \right)(y) \right| dy \leq c_p d^{l - \frac{n}{p}} \left\| \overline{D}^{\alpha} f \right\|_{L_p(B)}, \qquad x \in B. \qquad (2.23)
$$

Consequently, we derive

$$
\left| \sum_{|\alpha|=l} \int_B \frac{\left(\overline{D}^{\alpha} f \right)(y)}{|x - y|^{n-l}} w_{\alpha}(x, y) \, dy \right|
$$

$$
\leq \left(\sum_{|\alpha|=l} \left\| \overline{D}^{\alpha} f \right\|_{L_p(B)} \right) \| \omega \|_{L_{\infty}(\mathbb{R}^n)} n c_p d^{l - \frac{n}{p} + n}, \qquad x \in B. \qquad (2.24)
$$

Also we make

Remark 2.17. Again here $d = diamB$, assume $\omega \in C_0^\infty (\Omega)$, supp $\omega \subset \overline{B}$, i.e., $\|\omega\|_\infty < \infty$, $\int_{\mathbb{R}^n} \omega (x) \, dx = 1$. Here $\overline{D}^\alpha f$ is either $D^\alpha f$ or $D_w^\alpha f$. Then for $l \geq n + |\beta|$, $x \in B$, we obtain

$$\left| \sum_{|\alpha|=l, \alpha \geq \beta} \int_B \frac{\left(\overline{D}^\alpha f\right)(y)}{|x-y|^{n-l+|\beta|}} w_{\alpha-\beta}(x,y) \, dy \right|$$

$$\leq \left(\sum_{|\alpha|=l, \alpha \geq \beta} \left\| \overline{D}^\alpha f \right\|_{L_1(B)} \right) \|\omega\|_\infty \, n d^{l-|\beta|}. \qquad (2.25)$$

Also for $l \geq n + |\beta|$, $\left\| \overline{D}^\alpha f \right\|_{L_\infty(B)} < \infty$, $|\alpha| = l$, $\alpha \geq \beta$, $x \in B$, we get

$$\left| \sum_{|\alpha|=l, \alpha \geq \beta} \int_B \frac{\left(\overline{D}^\alpha f\right)(y)}{|x-y|^{n-l+|\beta|}} w_{\alpha-\beta}(x,y) \, dy \right|$$

$$\leq \left(\sum_{|\alpha|=l, \alpha \geq \beta} \left\| \overline{D}^\alpha f \right\|_{L_\infty(B)} \right) \|\omega\|_\infty \, \frac{n \pi^{\frac{n}{2}}}{2^n \Gamma \left(\frac{n}{2}+1\right)} d^{l-|\beta|+n}. \qquad (2.26)$$

Next, suppose $\overline{D}^\alpha f \in L_p(B)$, all $\alpha : |\alpha| = l$, $\alpha \geq \beta$, $1 < p < \infty$, $l > \frac{n}{p} + |\beta|$, $x \in B$. Then

$$\left| \sum_{|\alpha|=l, \alpha \geq \beta} \int_B \frac{\left(\overline{D}^\alpha f\right)(y)}{|x-y|^{n-l+|\beta|}} w_{\alpha-\beta}(x,y) \, dy \right|$$

$$\leq \left(\sum_{|\alpha|=l, \alpha \geq \beta} \left\| \overline{D}^\alpha f \right\|_{L_p(B)} \right) \|\omega\|_\infty \, n c_p d^{l-|\beta|-\frac{n}{p}+n}. \qquad (2.27)$$

We make

Remark 2.18. Here $\overline{D}^\alpha f$ denotes either $D^\alpha f$ or $D_w^\alpha f$, $d = diamB$. Suppose $\|\omega\|_{L_\infty(\mathbb{R}^n)} < \infty$. Denote by

$$Q^{l-1} f(x) := \sum_{1 \leq |\alpha| \leq l-1} \frac{1}{\alpha!} \int_B \left(\overline{D}^\alpha f\right)(y)(x-y)^\alpha \omega(y) \, dy, \quad \forall x \in \Omega, \quad (2.28)$$

the quasi-averaged Taylor polynomial. When $l = 1$, then $Q^0 f(x) := 0$.

In this chapter sums of the form $\displaystyle\sum_{1 \leq |\alpha| \leq 0} . = 0$.

Then for $x \in B$ we obtain

$$
\left| Q^{l-1} f(x) \right| \leq \left(\sum_{1 \leq |\alpha| \leq l-1} \left(\frac{1}{\alpha!} \int_B \left| \left(\overline{D}^\alpha f \right)(y) \right| \cdot |(x-y)^\alpha| \, dy \right) \right) \| \omega \|_{L_\infty(\mathbb{R}^n)}
$$

$$
\leq \left(\sum_{1 \leq |\alpha| \leq l-1} \left(\frac{1}{\alpha!} \int_B \left| \left(\overline{D}^\alpha f \right)(y) \right| \cdot |x-y|^{|\alpha|} \, dy \right) \right) \| \omega \|_{L_\infty(\mathbb{R}^n)}
$$

$$
=: (**). \tag{2.29}
$$

We notice that

$$
(**) \leq \left(\sum_{1 \leq |\alpha| \leq l-1} \left(\frac{d^{|\alpha|}}{\alpha!} \left(\int_B \left| \left(\overline{D}^\alpha f \right)(y) \right| dy \right) \right) \right) \| \omega \|_{L_\infty(\mathbb{R}^n)}
$$

$$
= \left\{ \sum_{1 \leq |\alpha| \leq l-1} \left(\frac{d^{|\alpha|}}{\alpha!} \left\| \overline{D}^\alpha f \right\|_{L_1(B)} \right) \right\} \| \omega \|_{L_\infty(\mathbb{R}^n)}.
$$

So we have proved for $x \in B$ that

$$
\left| Q^{l-1} f(x) \right| \leq \left\{ \sum_{1 \leq |\alpha| \leq l-1} \left(\frac{d^{|\alpha|}}{\alpha!} \left\| \overline{D}^\alpha f \right\|_{L_1(B)} \right) \right\} \| \omega \|_{L_\infty(\mathbb{R}^n)}. \tag{2.30}
$$

Also, when $\left\| \overline{D}^\alpha f \right\|_{L_\infty(B)} < \infty$, for all $\alpha : 1 \leq |\alpha| \leq l-1$, we get

$$
\left(\sum_{1 \leq |\alpha| \leq l-1} \left(\frac{d^{|\alpha|}}{\alpha!} \left(\int_B \left| \left(\overline{D}^\alpha f \right)(y) \right| dy \right) \right) \right) \| \omega \|_{L_\infty(\mathbb{R}^n)}
$$

$$
\leq \left(\sum_{1 \leq |\alpha| \leq l-1} \left(\frac{d^{|\alpha|}}{\alpha!} \left\| \overline{D}^\alpha f \right\|_{L_\infty(B)} \right) \right) Vol(B) \, \| \omega \|_{L_\infty(\mathbb{R}^n)}
$$

$$
= \left(\sum_{1 \leq |\alpha| \leq l-1} \left(\frac{d^{|\alpha|}}{\alpha!} \left\| \overline{D}^\alpha f \right\|_{L_\infty(B)} \right) \right) \| \omega \|_{L_\infty(\mathbb{R}^n)} \frac{\pi^{\frac{n}{2}}}{\Gamma\left(\frac{n}{2}+1\right)} \frac{d^n}{2^n}.
$$

So we proved, when $\left\| \overline{D}^\alpha f \right\|_{L_\infty(B)} < \infty$, for all $\alpha : 1 \leq |\alpha| \leq l-1$, $x \in B$, that

$$
\left| Q^{l-1} f(x) \right| \leq \left(\sum_{1 \leq |\alpha| \leq l-1} \left(\frac{d^{(n+|\alpha|)}}{\alpha!} \left\| \overline{D}^\alpha f \right\|_{L_\infty(B)} \right) \right) \frac{\| \omega \|_{L_\infty(\mathbb{R}^n)} \cdot \pi^{\frac{n}{2}}}{2^n \Gamma\left(\frac{n}{2}+1\right)}. \tag{2.31}
$$

We need

Lemma 2.19. *Let Ω^* be a region of $\mathbb{R}^n$ of finite diameter $d_* > 0$ and $f \in L_p(\Omega^*)$, $1 < p, q < \infty : \frac{1}{p} + \frac{1}{q} = 1$ and $m \in \mathbb{N}$, then*

$$\int_{\Omega^*} |x - z|^m |f(z)| \, dz \leq c_{q,m,n} d_*^{\left(m + \frac{n}{q}\right)} \|f\|_{L_p(\Omega^*)}, \quad \forall \, x \in \Omega^*. \tag{2.32}$$

Proof. We see that

$$\int_{\Omega^*} |x - z|^m |f(z)| \, dz \leq \left(\int_{\Omega^*} |x - z|^{mq} \, dz\right)^{\frac{1}{q}} \|f\|_{L_p(\Omega^*)}$$

(using polar coordinates)

$$\leq C_n \left(\int_0^{d_*} r^{mq+n-1} dr\right)^{\frac{1}{q}} \|f\|_{L_p(\Omega^*)}$$

$$= c_{q,m,n} d_*^{\frac{mq+n}{q}} \|f\|_{L_p(\Omega^*)} = c_{q,m,n} d_*^{\left(m + \frac{n}{q}\right)} \|f\|_{L_p(\Omega^*)}.$$

∎

Remark 2.20 (continuing from Remark 2.18). Let $p, q > 1 : \frac{1}{p} + \frac{1}{q} = 1$. Assume $\left(\overline{D}^\alpha f\right) \in L_p(B)$, for all $1 \leq |\alpha| \leq l - 1$, $x \in B$, then by Lemma 2.19 we obtain

$$\left|Q^{l-1} f(x)\right| \leq \left(\sum_{1 \leq |\alpha| \leq l-1} \left(\frac{1}{\alpha!} \int_B \left|\left(\overline{D}^\alpha f\right)(y)\right| \cdot |x - y|^{|\alpha|} \, dy\right)\right) \|\omega\|_{L_\infty(\mathbb{R}^n)}$$

$$\leq \left(\sum_{1 \leq |\alpha| \leq l-1} \left(\frac{1}{\alpha!} c_{q,|\alpha|,n} d^{\left(|\alpha| + \frac{n}{q}\right)} \left\|\overline{D}^\alpha f\right\|_{L_p(B)}\right)\right) \|\omega\|_{L_\infty(\mathbb{R}^n)}.$$

That is

$$\left|Q^{l-1} f(x)\right| \leq c_{q,l,n} \|\omega\|_{L_\infty(\mathbb{R}^n)} \left(\sum_{1 \leq |\alpha| \leq l-1} \left(\frac{\left\|\overline{D}^\alpha f\right\|_{L_p(B)} \cdot d^{\left(|\alpha| + \frac{n}{q}\right)}}{\alpha!}\right)\right),$$

$$x \in B. \tag{2.33}$$

Remark 2.21. For $x \in B$, we consider here $\omega \in C_0^\infty(\Omega)$, supp $\omega \subset \overline{B}$, $\int_{\mathbb{R}^n} \omega(x) \, dx = 1$; $f \in C^l(\Omega)$ or $f \in \left(W_1^l\right)^{\text{loc}}(\Omega)$. Here $\overline{D}^\alpha$ denotes any of D^α, D_w^α. We also consider

$$Q^{l-1} f(x) = \int_B \left(\sum_{1 \le |\alpha| \le l-1} \frac{(-1)^{|\alpha|}}{\alpha!} D_y^\alpha [(x-y)^\alpha \omega(y)] \right) f(y) \, dy$$

$$= \sum_{1 \le |\alpha| \le l-1} \frac{(-1)^{|\alpha|}}{\alpha!} \int_B D_y^\alpha [(x-y)^\alpha \omega(y)] f(y) \, dy. \qquad (2.34)$$

Hence

$$|Q^{l-1} f(x)| \le \sum_{1 \le |\alpha| \le l-1} \frac{1}{\alpha!} \int_B \left| D_y^\alpha [(x-y)^\alpha \omega(y)] \right| \cdot |f(y)| \, dy. \qquad (2.35)$$

We derive $\forall \, x \in B$,

$$|Q^{l-1} f(x)| \le \begin{cases} \left(\sum_{1 \le |\alpha| \le l-1} \frac{1}{\alpha!} \left\| D_y^\alpha [(x-y)^\alpha \omega(y)] \right\|_\infty \right) \|f\|_{L_1(B)}, \\[2ex] \left(\sum_{1 \le |\alpha| \le l-1} \frac{1}{\alpha!} \left\| D_y^\alpha [(x-y)^\alpha \omega(y)] \right\|_{L_1(B)} \right) \|f\|_{L_\infty(B)}, \\[1ex] \qquad\qquad \text{if } f \in L_\infty(B), \\[1ex] \text{when } p, q > 1 : \frac{1}{p} + \frac{1}{q} = 1, \text{ we have} \\[1ex] \left(\sum_{1 \le |\alpha| \le l-1} \frac{1}{\alpha!} \left\| D_y^\alpha [(x-y)^\alpha \omega(y)] \right\|_{L_q(B)} \right) \|f\|_{L_p(B)}, \\[1ex] \qquad\qquad \text{if } f \in L_p(B). \end{cases} \qquad (2.36)$$

Let $\beta \in \mathbb{Z}_+^n$ and $0 < |\beta| < l$.

We consider here

$$Q_\beta^{l-1} f(x) := \int_B \left(\sum_{1 \le |\alpha| \le l-|\beta|-1} \frac{(-1)^{|\alpha|+|\beta|}}{\alpha!} D_y^{\alpha+\beta} [(x-y)^\alpha \omega(y)] \right) f(y) \, dy$$

$$= \sum_{1 \le |\alpha| \le l-|\beta|-1} \frac{(-1)^{|\alpha|+|\beta|}}{\alpha!} \int_B D_y^{\alpha+\beta} [(x-y)^\alpha \omega(y)]$$

$$\times f(y) \, dy. \qquad (2.37)$$

When $l = |\beta| + 1$, then $Q_\beta^{l-1} f(x) := 0$. Hence

$$\left| Q_\beta^{l-1} f(x) \right| \le \sum_{1 \le |\alpha| \le l-|\beta|-1} \frac{1}{\alpha!} \int_B \left| D_y^{\alpha+\beta} [(x-y)^\alpha \omega(y)] \right| \cdot |f(y)| \, dy. \qquad (2.38)$$

We obtain $\forall\, x \in B$,

$$
\left| Q_{\beta}^{l-1} f(x) \right| \leq
\begin{cases}
\left(\displaystyle\sum_{1 \leq |\alpha| \leq l-|\beta|-1} \frac{1}{\alpha!} \left\| D_y^{\alpha+\beta} \left[(x-y)^\alpha \, \omega(y) \right] \right\|_\infty \right) \| f \|_{L_1(B)}, \\[2em]
\left(\displaystyle\sum_{1 \leq |\alpha| \leq l-|\beta|-1} \frac{1}{\alpha!} \left\| D_y^{\alpha+\beta} \left[(x-y)^\alpha \, \omega(y) \right] \right\|_{L_1(B)} \right) \| f \|_{L_\infty(B)}, \\[1em]
\qquad\qquad \text{if } f \in L_\infty(B), \\[1em]
\text{when } p, q > 1 : \frac{1}{p} + \frac{1}{q} = 1, \text{ we have} \\[1em]
\left(\displaystyle\sum_{1 \leq |\alpha| \leq l-|\beta|-1} \frac{1}{\alpha!} \left\| D_y^{\alpha+\beta} \left[(x-y)^\alpha \, \omega(y) \right] \right\|_{L_q(B)} \right) \| f \|_{L_p(B)}, \\[1em]
\qquad\qquad \text{if } f \in L_p(B).
\end{cases}
\tag{2.39}
$$

The final remark follows.

Remark 2.22. Here $\overline{D}^\alpha$ denotes D^α or D_w^α, and $\overline{D}^\beta$ means D^β or D_w^β. We rewrite (2.4), (2.7), (2.15), and (2.17). For $x \in \Omega$ we get

$$
f(x) = \int_B f(y) \, \omega(y) \, dy + Q^{l-1} f(x) + R^l f(x), \tag{2.40}
$$

where

$$
R^l f(x) := \sum_{|\alpha|=l} \int_{V_x} \frac{\overline{D}^\alpha f(y)}{|x-y|^{n-l}} w_\alpha(x,y) \, dy \tag{2.41}
$$

is equal to the remainders of ((2.4)and (2.44), respectively).

Also for $x \in \Omega$ we write (2.18) as follows:

$$
\left(\overline{D}^\beta f \right)(x) = (-1)^{|\beta|} \int_B (D^\beta \omega)(y) \, f(y) \, dy + Q_\beta^{l-1} f(x) + R_\beta^l f(x), \tag{2.42}
$$

where

$$
R_\beta^l f(x) := \sum_{|\alpha|=l, \alpha \geq \beta} \int_{V_x} \frac{\left(\overline{D}^\alpha f \right)(y)}{|x-y|^{n-l+|\beta|}} w_{\alpha-\beta}(x,y) \, dy. \tag{2.43}
$$

Additionally we give

Theorem 2.23. *Let* $\Omega \subset \mathbb{R}^n$ *be a domain star-shaped with respect to the open ball* $B = B(x_0, r)$ *such that* $\overline{B} \subset \Omega$, $\omega \in L_\infty(\mathbb{R}^n)$, *supp* $\omega \subset \overline{B}$, $\int_{\mathbb{R}^n} \omega(x)\,dx = 1$, $l \in \mathbb{N}$ *and* $f \in \left(W_1^l\right)^{\text{loc}}(\Omega)$. *Then for almost every* $x \in \Omega$

$$f(x) = \sum_{|\alpha|<l} \frac{1}{\alpha!} \int_B \left(D_w^\alpha f\right)(y)(x-y)^\alpha \omega(y)\,dy$$

$$+ l \sum_{|\alpha|=l} \frac{1}{\alpha!} \int_B (x-y)^\alpha \omega(y)$$

$$\times \left(\int_0^1 (1-t)^{l-1}\left(D_w^\alpha f\right)(y + t(x-y))\,dt\right) dy. \qquad (2.44)$$

Proof. From the assumptions of the theorem, we get for almost every $x \in \Omega$ that

$$f(x) - \sum_{|\alpha|<l} \frac{1}{\alpha!} \int_B \left(D_w^\alpha f\right)(y)(x-y)^\alpha \omega(y)\,dy$$

$$= \sum_{|\alpha|=l} \int_{V_x} \frac{\left(D_w^\alpha f\right)(y)}{|x-y|^{n-l}} w_\alpha(x, y)\,dy,$$

implying that $\int_{V_x} \frac{(D_w^\alpha f)(y)}{|x-y|^{n-l}} w_\alpha(x, y)\,dy$ is finite for almost every $x \in \Omega$.

From [6], p. 105, (3.41) there, we know that $\forall\, x \in \mathbb{R}^n$

$$\sup p_y w_\alpha(x, y) = \sup p_y w(x, y) \subset \overline{K_x},$$

where K_x is the cone in $\mathbb{R}^n$ related to V_x, see again [6], pp. 93–100.

So acting similarly to [6], p. 107 and working backwards we derive

$$\sum_{|\alpha|=l} \int_{V_x} \frac{\left(D_w^\alpha f\right)(y)}{|x-y|^{n-l}} w_\alpha(x, y)\,dy = l \sum_{|\alpha|=l} \frac{1}{\alpha!} J_\alpha,$$

where

$$J_\alpha = \int_{\mathbb{R}^n} \left(D_w^\alpha f\right)(z) \frac{(x-z)^\alpha}{|x-z|^n} \left(\int_{|x-z|}^\infty \omega\left(x + \rho \frac{z-x}{|z-x|}\right) \rho^{n-1}\,d\rho\right) dz.$$

Replacing ρ by $\frac{|x-z|}{1-t}$, we obtain

$$J_\alpha = \int_{\mathbb{R}^n} \left(D_w^\alpha f\right)(z)(x-z)^\alpha \left(\int_0^1 \omega\left(\frac{z-tx}{1-t}\right) \frac{dt}{(1-t)^{n+1}}\right) dz.$$

Next, setting $z = y + t(x - y)$ and noticing $(x - z)^\alpha = (1 - t)^l (x - y)^\alpha$ and $dz = (1 - t)^n dy$ we find that

$$
\begin{aligned}
J_\alpha &= \int_0^1 (1 - t)^{l-1} \left(\int_{\mathbb{R}^n} (D_w^\alpha f)(y + t(x - y))(x - y)^\alpha \, \omega(y) \, dy \right) dt \\
&= \int_{\mathbb{R}^n} (x - y)^\alpha \, \omega(y) \left(\int_0^1 (1 - t)^{l-1} (D_w^\alpha f)(y + t(x - y)) \, dt \right) dy \\
&= \int_B (x - y)^\alpha \, \omega(y) \left(\int_0^1 (1 - t)^{l-1} (D_w^\alpha f)(y + t(x - y)) \, dt \right) dy.
\end{aligned}
$$

We have proved that

$$
\sum_{|\alpha|=l} \int_{V_x} \frac{(D_w^\alpha f)(y)}{|x - y|^{n-l}} w_\alpha(x, y) \, dy
$$

$$
= l \sum_{|\alpha|=l} \frac{1}{\alpha!} \int_B (x - y)^\alpha \, \omega(y) \left(\int_0^1 (1 - t)^{l-1} (D_w^\alpha f)(y + t(x - y)) \, dt \right) dy,
$$

establishing the claim. ∎

Proposition 2.24. *Same assumptions as in Theorem 2.23. Then for almost every* $x \in B$ *we get*

$$
|Remainder\ (2.44)| \leq l d^l \, \|\omega\|_{L_\infty(\mathbb{R}^n)} \sum_{|\alpha|=l} \frac{1}{\alpha!} \left\| (D_w^\alpha f)(y + t(x - y)) \right\|_{L_1(B \times [0,1])}.
$$

$$(2.45)$$

In (2.45) we assume for all $\alpha : |\alpha| = l$ *that*

$$
\left\| (D_w^\alpha f)(y + t(x - y)) \right\|_{L_1(B \times [0,1])} < \infty.
$$

Proof. We have that

$$
\left| l \sum_{|\alpha|=l} \frac{1}{\alpha!} \int_B (x - y)^\alpha \, \omega(y) \left(\int_0^1 (1 - t)^{l-1} (D_w^\alpha f)(y + t(x - y)) \, dt \right) dy \right|
$$

$$
\leq l \sum_{|\alpha|=l} \frac{1}{\alpha!} \int_B |(x - y)^\alpha| \cdot |\omega(y)|
$$

$$
\times \left(\int_0^1 (1 - t)^{l-1} |(D_w^\alpha f)(y + t(x - y))| \, dt \right) dy
$$

$$\leq l d^l \, \|\omega\|_{L_\infty(\mathbb{R}^n)} \sum_{|\alpha|=l} \frac{1}{\alpha!} \int_B \left(\int_0^1 (1-t)^{l-1} \left| (D_w^\alpha f)(y+t(x-y)) \right| dt \right) dy$$

$$\leq l d^l \, \|\omega\|_{L_\infty(\mathbb{R}^n)} \sum_{|\alpha|=l} \frac{1}{\alpha!} \int_B \left(\int_0^1 \left| (D_w^\alpha f)(y+t(x-y)) \right| dt \right) dy$$

$$= l d^l \, \|\omega\|_{L_\infty(\mathbb{R}^n)} \sum_{|\alpha|=l} \frac{1}{\alpha!} \left\| (D_w^\alpha f)(y+t(x-y)) \right\|_{L_1(B \times [0,1])},$$

proving the claim. ∎

2.3 Main Results

On the way to prove the general Chebyshev–Grüss-type inequalities, we establish
the general

Theorem 2.25. *For f, g under the assumptions of any of Theorems 2.6, 2.8, 2.11, 2.23 and Corollary 2.12 we obtain that*

$$\Delta(f,g) := \left| \int_B \omega(x) f(x) g(x) \, dx - \left(\int_B \omega(x) f(x) \, dx \right) \left(\int_B \omega(x) g(x) \, dx \right) \right|$$

$$\leq \frac{1}{2} \left[\left(\int_B |\omega(x)| \, |g(x)| \, |Q^{l-1} f(x)| \, dx \right. \right.$$

$$+ \int_B |\omega(x)| \, |f(x)| \, |Q^{l-1} g(x)| \, dx \bigg)$$

$$+ \left(\int_B |\omega(x)| \, |g(x)| \, |R^l f(x)| \, dx \right.$$

$$+ \int_B |\omega(x)| \, |f(x)| \, |R^l g(x)| \, dx \bigg) \bigg]. \tag{2.46}$$

Proof. For $x \in B$ we have

$$f(x) = \int_B f(y) \omega(y) \, dy + Q^{l-1} f(x) + R^l f(x),$$

and

$$g(x) = \int_B g(y) \omega(y) \, dy + Q^{l-1} g(x) + R^l g(x).$$

Hence

$$\omega\,(x)\,f\,(x)\,g\,(x)$$
$$= \omega\,(x)\,g\,(x)\int_B f\,(y)\,\omega\,(y)\,\mathrm{d}y + \omega\,(x)\,g\,(x)\,Q^{l-1}f\,(x) + \omega\,(x)\,g\,(x)\,R^l\,f\,(x),$$

and

$$\omega\,(x)\,f\,(x)\,g\,(x)$$
$$= \omega\,(x)\,f\,(x)\int_B g\,(y)\,\omega\,(y)\,\mathrm{d}y + \omega\,(x)\,f\,(x)\,Q^{l-1}g\,(x) + \omega\,(x)\,f\,(x)\,R^l g\,(x).$$

Therefore

$$\int_B \omega\,(x)\,f\,(x)\,g\,(x)\,\mathrm{d}x = \left(\int_B \omega\,(x)\,g\,(x)\,\mathrm{d}x\right)\left(\int_B f\,(y)\,\omega\,(y)\,\mathrm{d}y\right)$$
$$+ \int_B \omega\,(x)\,g\,(x)\,Q^{l-1}f\,(x)\,\mathrm{d}x$$
$$+ \int_B \omega\,(x)\,g\,(x)\,R^l\,f\,(x)\,\mathrm{d}x,$$

and

$$\int_B \omega\,(x)\,f\,(x)\,g\,(x)\,\mathrm{d}x = \left(\int_B \omega\,(x)\,f\,(x)\,\mathrm{d}x\right)\left(\int_B g\,(y)\,\omega\,(y)\,\mathrm{d}y\right)$$
$$+ \int_B \omega\,(x)\,f\,(x)\,Q^{l-1}g\,(x)\,\mathrm{d}x$$
$$+ \int_B \omega\,(x)\,f\,(x)\,R^l g\,(x)\,\mathrm{d}x.$$

Consequently, there hold

$$\int_B \omega\,(x)\,f\,(x)\,g\,(x)\,\mathrm{d}x - \left(\int_B \omega\,(x)\,g\,(x)\,\mathrm{d}x\right)\left(\int_B f\,(x)\,\omega\,(x)\,\mathrm{d}x\right)$$
$$= \int_B \omega\,(x)\,g\,(x)\,Q^{l-1}f\,(x)\,\mathrm{d}x + \int_B \omega\,(x)\,g\,(x)\,R^l\,f\,(x)\,\mathrm{d}x,$$

and

$$\int_B \omega\,(x)\,f\,(x)\,g\,(x)\,\mathrm{d}x - \left(\int_B \omega\,(x)\,f\,(x)\,\mathrm{d}x\right)\left(\int_B g\,(x)\,\omega\,(x)\,\mathrm{d}x\right)$$
$$= \int_B \omega\,(x)\,f\,(x)\,Q^{l-1}g\,(x)\,\mathrm{d}x + \int_B \omega\,(x)\,f\,(x)\,R^l g\,(x)\,\mathrm{d}x.$$

Adding the last two equalities we obtain

$$
\int_B \omega(x)\, f(x)\, g(x)\, dx - \left(\int_B \omega(x)\, f(x)\, dx\right)\left(\int_B \omega(x)\, g(x)\, dx\right)
$$

$$
= \frac{1}{2}\left[\left(\int_B \omega(x)\, g(x)\, Q^{l-1} f(x)\, dx + \int_B \omega(x)\, f(x)\, Q^{l-1} g(x)\, dx\right)\right.
$$

$$
\left. + \left(\int_B \omega(x)\, g(x)\, R^l f(x)\, dx + \int_B \omega(x)\, f(x)\, R^l g(x)\, dx\right)\right],
$$

hence proving the claim.　∎

We give

Theorem 2.26. *Let $\Omega \subset \mathbb{R}^n$ be a domain star-shaped with respect to the open ball $B = B(x_0, r)$ such that $\overline{B} \subset \Omega$, $\omega \in L_1(\mathbb{R}^n)$, supp $\omega \subset \overline{B}$, $\int_{\mathbb{R}^n} \omega(x)\, dx = 1$, $l \in \mathbb{N}$ and $f, g \in C^l(\Omega)$. Then*

$$
\left|\int_B \omega(x)\, f(x)\, g(x)\, dx - \left(\int_B \omega(x)\, f(x)\, dx\right)\left(\int_B \omega(x)\, g(x)\, dx\right)\right|
$$

$$
\leq \frac{\|\omega\|^2_{L_1(\mathbb{R}^n)}}{2}\left[\left[\|g\|_{\infty,B}\left(\sum_{1 \leq |\alpha| \leq l-1} \frac{d^{|\alpha|}\, \|D^\alpha f\|_{\infty,B}}{\alpha!}\right)\right.\right.
$$

$$
\left. + \|f\|_{\infty,B}\left(\sum_{1 \leq |\alpha| \leq l-1} \frac{d^{|\alpha|}\, \|D^\alpha g\|_{\infty,B}}{\alpha!}\right)\right]
$$

$$
\left.+ \left[\frac{(nd)^l}{l!}\left(\|g\|_{\infty,B}\, \|D^\alpha f\|^{\max}_{\infty,l,B} + \|f\|_{\infty,B}\, \|D^\alpha g\|^{\max}_{\infty,l,B}\right)\right]\right],
$$

$$
(2.47)
$$

where d is the diameter of B.

When $l = 1$ the sums in (2.47) collapse.

Proof. One also in general obtains $(x \in B)$

$$
\left|Q^{l-1} f(x)\right| \leq \sum_{1 \leq |\alpha| \leq l-1} \frac{1}{\alpha!}\int_B \left|\overline{D}^\alpha f(y)\right| \cdot |(x-y)^\alpha|\, |\omega(y)|\, dy
$$

$$
\leq \sum_{1 \leq |\alpha| \leq l-1} \frac{1}{\alpha!}\left(\int_B |\omega(y)|\, dy\right) d^{|\alpha|}\left\|\overline{D}^\alpha f\right\|_{L_\infty(B)}
$$

$$
= \sum_{1 \leq |\alpha| \leq l-1} \frac{1}{\alpha!}\, \|\omega\|_{L_1(B)}\, d^{|\alpha|}\left\|\overline{D}^\alpha f\right\|_{L_\infty(B)}.
$$

So for $\left\| \overline{D}^\alpha f \right\|_{L_\infty(B)} < \infty$, for all $\alpha : 1 \leq |\alpha| \leq l - 1$, we proved

$$\left| Q^{l-1} f(x) \right| \leq \|\omega\|_{L_1(B)} \left(\sum_{1 \leq |\alpha| \leq l-1} \frac{d^{|\alpha|} \left\| \overline{D}^\alpha f \right\|_{L_\infty(B)}}{\alpha!} \right), \tag{2.48}$$

for $x \in B$.

By (2.46), Proposition 2.7 and (2.48) we obtain

$$\begin{aligned}
\Delta(f, g) &\leq \frac{1}{2} \Bigg[\|\omega\|_{L_1(B)}^2 \Bigg[\|g\|_{\infty,B} \left(\sum_{1 \leq |\alpha| \leq l-1} \frac{d^{|\alpha|} \|D^\alpha f\|_{\infty,B}}{\alpha!} \right) \\
&\qquad + \|f\|_{\infty,B} \left(\sum_{1 \leq |\alpha| \leq l-1} \frac{d^{|\alpha|} \|D^\alpha g\|_{\infty,B}}{\alpha!} \right) \Bigg] \\
&\qquad + \|\omega\|_{L_1(B)} \Bigg[\|g\|_{\infty,B} \frac{(nd)^l \|\omega\|_{L_1(\mathbb{R}^n)} \|D^\alpha f\|_{\infty,l,B}^{\max}}{l!} \\
&\qquad + \|f\|_{\infty,B} \frac{(nd)^l \|\omega\|_{L_1(\mathbb{R}^n)} \|D^\alpha g\|_{\infty,l,B}^{\max}}{l!} \Bigg] \Bigg] \\
&= \frac{\|\omega\|_{L_1(\mathbb{R}^n)}^2}{2} \Bigg[\Bigg[\|g\|_{\infty,B} \left(\sum_{1 \leq |\alpha| \leq l-1} \frac{d^{|\alpha|} \|D^\alpha f\|_{\infty,B}}{\alpha!} \right) \\
&\qquad + \|f\|_{\infty,B} \left(\sum_{1 \leq |\alpha| \leq l-1} \frac{d^{|\alpha|} \|D^\alpha g\|_{\infty,B}}{\alpha!} \right) \Bigg] \\
&\qquad + \Bigg[\frac{(nd)^l}{l!} \left(\|g\|_{\infty,B} \|D^\alpha f\|_{\infty,l,B}^{\max} + \|f\|_{\infty,B} \|D^\alpha g\|_{\infty,l,B}^{\max} \right) \Bigg] \Bigg],
\end{aligned}$$

proving the claim. ∎

We present

Theorem 2.27. *Let $\Omega \subset \mathbb{R}^n$ be a domain star-shaped with respect to the open ball $B = B(x_0, r)$ such that $\overline{B} \subset \Omega$, $\omega \in L_\infty(\mathbb{R}^n)$, $\mathrm{supp}\, \omega \subset \overline{B}$, $\int_{\mathbb{R}^n} \omega(x)\, dx = 1$, $l \in \mathbb{N}$ and $f, g \in \left(W_1^l \right)^{\mathrm{loc}} (\Omega)$. Suppose further that $l \geq n$. Then*

$$\left| \int_B \omega(x) f(x) g(x) \, dx - \left(\int_B \omega(x) f(x) \, dx \right) \left(\int_B \omega(x) g(x) \, dx \right) \right|$$

$$\leq \frac{\|\omega\|_{L_\infty(\mathbb{R}^n)}^2}{2} \left[\left\{ \|g\|_{L_1(B)} \left(\sum_{1 \leq |\alpha| \leq l-1} \left(\frac{d^{|\alpha|}}{\alpha!} \|D_w^\alpha f\|_{L_1(B)} \right) \right) \right. \right.$$

$$+ \|f\|_{L_1(B)} \left(\sum_{1 \leq |\alpha| \leq l-1} \left(\frac{d^{|\alpha|}}{\alpha!} \|D_w^\alpha g\|_{L_1(B)} \right) \right) \right\}$$

$$+ (nd)^l \left[\|g\|_{L_1(B)} \left(\sum_{|\alpha|=l} \|D_w^\alpha f\|_{L_1(B)} \right) \right.$$

$$\left. \left. + \|f\|_{L_1(B)} \left(\sum_{|\alpha|=l} \|D_w^\alpha g\|_{L_1(B)} \right) \right] \right]. \quad (2.49)$$

Proof. Here we get by (2.43), (2.30), and (2.20) that

$$\Delta(f,g) \leq \frac{\|\omega\|_{L_\infty(\mathbb{R}^n)}^2}{2} \left[\left\{ \|g\|_{L_1(B)} \left(\sum_{1 \leq |\alpha| \leq l-1} \left(\frac{d^{|\alpha|}}{\alpha!} \|D_w^\alpha f\|_{L_1(B)} \right) \right) \right. \right.$$

$$+ \|f\|_{L_1(B)} \left(\sum_{1 \leq |\alpha| \leq l-1} \left(\frac{d^{|\alpha|}}{\alpha!} \|D_w^\alpha g\|_{L_1(B)} \right) \right) \right\}$$

$$+ \left\{ \left[\|g\|_{L_1(B)} \left(\sum_{|\alpha|=l} \|D_w^\alpha f\|_{L_1(B)} \right) \right. \right.$$

$$\left. \left. \left. + \|f\|_{L_1(B)} \left(\sum_{|\alpha|=l} \|D_w^\alpha g\|_{L_1(B)} \right) \right] nd^l \right\} \right],$$

proving the claim. ∎

Based on (2.46), (2.31), and (2.21) we have

Theorem 2.28. *Let $\Omega \subset \mathbb{R}^n$ be a domain star-shaped with respect to the open ball $B = B(x_0, r)$ such that $\overline{B} \subset \Omega$, $\omega \in L_\infty(\mathbb{R}^n)$, supp $\omega \subset \overline{B}$,*

$\int_{\mathbb{R}^n} \omega(x)\,dx = 1$, $l \in \mathbb{N}$ and $f, g \in \left(W_1^l\right)^{\text{loc}}(\Omega)$. *Furthermore assume* $\left\|D_w^\alpha f\right\|_{L_\infty(B)}, \left\|D_w^\alpha g\right\|_{L_\infty(B)} < \infty$ *for all* $\alpha : 1 \leq |\alpha| \leq l, l \geq n$. *Then*

$$\left| \int_B \omega(x) f(x) g(x)\,dx - \left(\int_B \omega(x) f(x)\,dx \right) \left(\int_B \omega(x) g(x)\,dx \right) \right|$$

$$\leq \frac{\pi^{\frac{n}{2}} \|\omega\|_{L_\infty(\mathbb{R}^n)}^2}{2^{n+1}\Gamma\left(\frac{n}{2}+1\right)} \left\{ \left[\|g\|_{L_1(B)} \left(\sum_{1 \leq |\alpha| \leq l-1} \left(\frac{d^{n+|\alpha|}}{\alpha!} \left\|D_w^\alpha f\right\|_{L_\infty(B)} \right) \right) \right. \right.$$

$$\left. + \|f\|_{L_1(B)} \left(\sum_{1 \leq |\alpha| \leq l-1} \left(\frac{d^{n+|\alpha|}}{\alpha!} \left\|D_w^\alpha g\right\|_{L_\infty(B)} \right) \right) \right]$$

$$+ \left[\|g\|_{L_1(B)} \left(\sum_{|\alpha|=l} \left\|D_w^\alpha f\right\|_{L_\infty(B)} \right) \right.$$

$$\left. \left. + \|f\|_{L_1(B)} \left(\sum_{|\alpha|=l} \left\|D_w^\alpha g\right\|_{L_\infty(B)} \right) \right] nd^{l+n} \right\}. \quad (2.50)$$

Based on (2.46), (2.33), and (2.24) we get

Theorem 2.29. *Let* $\Omega \subset \mathbb{R}^n$ *be a domain star-shaped with respect to the open ball* $B = B(x_0, r)$ *such that* $\overline{B} \subset \Omega$, $\omega \in L_\infty(\mathbb{R}^n)$, *supp* $\omega \subset \overline{B}$, $\int_{\mathbb{R}^n} \omega(x)\,dx = 1$, $l \in \mathbb{N}$ *and* $f, g \in \left(W_1^l\right)^{\text{loc}}(\Omega)$. *Furthermore suppose* $p, q > 1 : \frac{1}{p} + \frac{1}{q} = 1, l > \frac{n}{p}$, *for all* $\alpha : 1 \leq |\alpha| \leq l$, $D_w^\alpha f, D_w^\alpha g \in L_p(B)$. *Then*

$$\left| \int_B \omega(x) f(x) g(x)\,dx - \left(\int_B \omega(x) f(x)\,dx \right) \left(\int_B \omega(x) g(x)\,dx \right) \right|$$

$$\leq \frac{\|\omega\|_{L_\infty(\mathbb{R}^n)}^2}{2} \left\{ \left[\|g\|_{L_1(B)} \left(\sum_{|\alpha|=l} \left\|D_w^\alpha f\right\|_{L_p(B)} \right) \right. \right.$$

$$\left. + \|f\|_{L_1(B)} \left(\sum_{|\alpha|=l} \left\|D_w^\alpha g\right\|_{L_p(B)} \right) \right] nc_p d^{l-\frac{n}{p}+n}$$

$$+ c_{q,l,n} \left[\|g\|_{L_1(B)} \left(\sum_{1 \leq |\alpha| \leq l-1} \left(\frac{\left\|D_w^\alpha f\right\|_{L_p(B)} d^{|\alpha|+\frac{n}{q}}}{\alpha!} \right) \right) \right.$$

$$\left. \left. + \|f\|_{L_1(B)} \left(\sum_{1 \leq |\alpha| \leq l-1} \left(\frac{\left\|D_w^\alpha g\right\|_{L_p(B)} d^{|\alpha|+\frac{n}{q}}}{\alpha!} \right) \right) \right] \right\}.$$

$$(2.51)$$

Remark 2.30. When $f, g \in C^l(\Omega)$ the Theorems 2.27, 2.28, 2.29 are again valid. In this case we replace D_w^α by D_w in all inequalities (2.49), (2.50), and (2.51).

We give

Theorem 2.31. *Let $\Omega \subset \mathbb{R}^n$ be a domain star-shaped with respect to the open ball $B = B(x_0, r)$ such that $\overline{B} \subset \Omega$, $\omega \in C_0^\infty(\Omega)$, supp $\omega \subset \overline{B}$, $\int_{\mathbb{R}^n} \omega(x) \, dx = 1$. Let f, g either in $C^l(\Omega)$ or in $\left(W_1^l\right)^{loc}(\Omega)$. Here $\overline{D}^\alpha f$ denotes either $D^\alpha f$ or $D_w^\alpha f$ and $\Delta(f, g)$ is as in (2.46).*

We have the following cases:

(i) Here $l \geq n$. Then

$$\Delta(f, g) \leq \frac{\|\omega\|_\infty}{2} \left[\left\{ 2 \|g\|_{L_1(B)} \|f\|_{L_1(B)} \right. \right.$$

$$\left. \times \left(\sum_{1 \leq |\alpha| \leq l-1} \frac{1}{\alpha!} \left\| D_y^\alpha \left[(x - y)^\alpha \, w(y) \right] \right\|_{\infty, B^2} \right) \right\}$$

$$+ \|\omega\|_\infty \, n d^l \left[\|g\|_{L_1(B)} \left(\sum_{|\alpha|=l} \left\| \overline{D}^\alpha f \right\|_{L_1(B)} \right) \right.$$

$$\left. \left. + \|f\|_{L_1(B)} \left(\sum_{|\alpha|=l} \left\| \overline{D}^\alpha g \right\|_{L_1(B)} \right) \right] \right]. \qquad (2.52)$$

(ii) Here $l \geq n$; $f, g \in L_\infty(B)$ and $\overline{D}^\alpha f, \overline{D}^\alpha g \in L_\infty(B)$ for all $\alpha : |\alpha| = l$. Then

$$\Delta(f, g) \leq \frac{\|\omega\|_\infty}{2} \left\{ \left(\|f\|_{L_\infty(B)} \|g\|_{L_1(B)} + \|f\|_{L_1(B)} \|g\|_{L_\infty(B)} \right) \right.$$

$$\times \left(\sum_{1 \leq |\alpha| \leq l-1} \frac{1}{\alpha!} \sup_{x \in B} \left\| D_y^\alpha \left[(x - y)^\alpha \, w(y) \right] \right\|_{L_1(B)} \right)$$

$$+ \frac{n d^{l+n} \pi^{\frac{n}{2}} \|\omega\|_\infty}{2^n \Gamma\left(\frac{n}{2} + 1\right)} \left[\|g\|_{L_1(B)} \left(\sum_{|\alpha|=l} \left\| \overline{D}^\alpha f \right\|_{L_\infty(B)} \right) \right.$$

$$\left. \left. + \|f\|_{L_1(B)} \left(\sum_{|\alpha|=l} \left\| \overline{D}^\alpha g \right\|_{L_\infty(B)} \right) \right] \right\}. \qquad (2.53)$$

*(iii) Let $p, q > 1 : \frac{1}{p} + \frac{1}{q} = 1, l > \frac{n}{p}$, for all $\alpha : |\alpha| = l, \overline{D}^\alpha f, \overline{D}^\alpha g, f, g \in$
$L_p(B)$. Then*

$$\Delta(f, g) \le \frac{\|\omega\|_\infty}{2} \left[\left(\|g\|_{L_1(B)} \|f\|_{L_p(B)} + \|f\|_{L_1(B)} \|g\|_{L_p(B)} \right) \cdot \right.$$

$$\left(\sum_{1 \le |\alpha| \le l-1} \frac{1}{\alpha!} \sup_{x \in B} \left\| D_y^\alpha [(x-y)^\alpha w(y)] \right\|_{L_q(B)} \right)$$

$$+ \|\omega\|_\infty c_{q,l,n} \left[\|g\|_{L_1(B)} \left(\sum_{1 \le |\alpha| \le l-1} \left(\frac{\|\overline{D}^\alpha f\|_{L_p(B)}}{\alpha!} d^{|\alpha| + \frac{n}{q}} \right) \right) \right.$$

$$\left. \left. + \|f\|_{L_1(B)} \left(\sum_{1 \le |\alpha| \le l-1} \left(\frac{\|\overline{D}^\alpha g\|_{L_p(B)}}{\alpha!} d^{|\alpha| + \frac{n}{q}} \right) \right) \right] \right]. \quad (2.54)$$

Proof. By use of (2.36) and Theorems 2.27–2.29. ■

We also present

Theorem 2.32. *Let $\Omega \subset \mathbb{R}^n$ be a domain star-shaped with respect to the open ball $B = B(x_0, r)$ such that $\overline{B} \subset \Omega$, $\omega \in L_\infty(\mathbb{R}^n)$, supp $\omega \subset \overline{B}$, $\int_{\mathbb{R}^n} \omega(x) dx = 1$, $l \in \mathbb{N}$ and $f, g \in \left(W_1^l \right)^{loc}(\Omega)$. Furthermore suppose for all $\alpha : |\alpha| = l \ge n$ that $\left\| (D_w^\alpha f)(y + t(x-y)) \right\|_{L_1(B \times [0,1])}, \left\| (D_w^\alpha g)(y + t(x-y)) \right\|_{L_1(B \times [0,1])} < \infty$. Then*

$$\Delta(f, g) \le \frac{\|\omega\|_{L_\infty(\mathbb{R}^n)}^2}{2} \cdot \left[\left\{ \|g\|_{L_1(B)} \left(\sum_{1 \le |\alpha| \le l-1} \left(\frac{d^{|\alpha|}}{\alpha!} \|D_w^\alpha f\|_{L_1(B)} \right) \right) \right. \right.$$

$$\left. + \|f\|_{L_1(B)} \left(\sum_{1 \le |\alpha| \le l-1} \left(\frac{d^{|\alpha|}}{\alpha!} \|D_w^\alpha g\|_{L_1(B)} \right) \right) \right\}$$

$$+ ld^l \left[\|g\|_{L_1(B)} \left(\sum_{|\alpha| = l} \frac{1}{\alpha!} \|(D_w^\alpha f)(y + t(x-y))\|_{L_1(B \times [0,1])} \right) \right.$$

$$\left. \left. + \|f\|_{L_1(B)} \left(\sum_{|\alpha| = l} \frac{1}{\alpha!} \|(D_w^\alpha g)(y + t(x-y))\|_{L_1(B \times [0,1])} \right) \right] \right]$$

$$\quad (2.55)$$

Proof. By Theorem 2.27 and Proposition 2.24. ■

Next, we give a series of Ostrowski-type inequalities.

Theorem 2.33. *Let all as in Theorem 2.6. Call again*

$$Q^{l-1} f(x) := \sum_{1 \le |\alpha| \le l-1} \frac{1}{\alpha!} \int_B (D^\alpha f)(y)(x-y)^\alpha \omega(y)\,dy.$$

Then for every $x \in B$, we obtain

$$\left| f(x) - \int_B f(y)\omega(y)\,dy - Q^{l-1} f(x) \right|$$

$$\le \frac{(nd)^l \|\omega\|_{L_1(\mathbb{R}^n)} \|D^\alpha f\|_{\infty,l,B}^{\max}}{l!} := A_1. \tag{2.56}$$

Also it holds, by additionally assuming $\|\omega\|_{L_\infty(\mathbb{R}^n)} < \infty$, that

$$\left| f(x) - \int_B f(y)\omega(y)\,dy \right|$$

$$\le \left(\sum_{1 \le |\alpha| \le l-1} \left(\frac{d^{n+|\alpha|}}{\alpha!} \|D^\alpha f\|_\infty \right) \right) \frac{\|\omega\|_{L_\infty(\mathbb{R}^n)} \pi^{\frac{n}{2}}}{2^n \Gamma\left(\frac{n}{2}+1\right)}$$

$$+ \frac{(nd)^l \|\omega\|_{L_1(\mathbb{R}^n)} \|D^\alpha f\|_{\infty,l,B}^{\max}}{l!} := B_1. \tag{2.57}$$

Proof. Use of Theorem 2.6, (2.40), (2.5), and (2.31). ∎

We continue with

Theorem 2.34. *All as in Theorems 2.8 or 2.11. Assume $l \ge n$, $\|\omega\|_{L_\infty(\mathbb{R}^n)} < \infty$. Then for every $x \in B$ (almost every $x \in B$, respectively) we get*

$$E(f)(x) := \left| f(x) - \int_B f(y)\omega(y)\,dy - Q^{l-1} f(x) \right|$$

$$\le \left(\sum_{|\alpha|=l} \left\| \overline{D}^\alpha f \right\|_{L_1(B)} \right) \|\omega\|_{L_\infty(\mathbb{R}^n)} nd^l =: A_2. \tag{2.58}$$

Also it holds

$$\Delta(f)(x) := \left| f(x) - \int_B f(y)\omega(y)\,dy \right|$$

$$\le \left[\left\{ \sum_{1 \le |\alpha| \le l-1} \left(\frac{d^{|\alpha|}}{\alpha!} \left\| \overline{D}^\alpha f \right\|_{L_1(B)} \right) \right\} \right.$$

$$\left. + \left(\sum_{|\alpha|=l} \left\| \overline{D}^\alpha f \right\|_{L_1(B)} \right) nd^l \right] \|\omega\|_{L_\infty(\mathbb{R}^n)} =: B_2. \tag{2.59}$$

Proof. Use of Theorems 2.8, 2.11; (2.40), (2.20), (2.28), and (2.30).   ∎

We give

Theorem 2.35. *All as in Theorems 2.8, 2.11. Suppose $l \geq n$, $\|\omega\|_{L_\infty(\mathbb{R}^n)} < \infty$, and $\left\|\overline{D}^\alpha f\right\|_{L_\infty(B)} < \infty$ for all $\alpha : |\alpha| = l$. Then for every $x \in B$ (almost every $x \in B$, respectively), it holds*

$$E(f)(x) \leq \left(\sum_{|\alpha|=l} \left\|\overline{D}^\alpha f\right\|_{L_\infty(B)} \right) \|\omega\|_{L_\infty(\mathbb{R}^n)} \frac{n\pi^{\frac{n}{2}}}{2^n \Gamma\left(\frac{n}{2}+1\right)} d^{l+n} =: A_3. \quad (2.60)$$

Additionally, assume that $\left\|\overline{D}^\alpha f\right\|_{L_\infty(B)} < \infty$ for all $\alpha : 1 \leq |\alpha| \leq l - 1$. It holds

$$\Delta(f)(x) \leq \left[\left(\sum_{1 \leq |\alpha| \leq l-1} \left(\frac{d^{n+|\alpha|}}{\alpha!} \left\|\overline{D}^\alpha f\right\|_{L_\infty(B)} \right) \right) \right.$$

$$\left. + \left(\sum_{|\alpha|=l} \left\|\overline{D}^\alpha f\right\|_{L_\infty(B)} \right) n d^{l+n} \right] \frac{\pi^{\frac{n}{2}}}{2^n \Gamma\left(\frac{n}{2}+1\right)} \|\omega\|_{L_\infty(\mathbb{R}^n)}$$

$$=: B_3. \qquad\qquad\qquad\qquad\qquad\qquad\qquad\qquad\qquad (2.61)$$

Proof. Use of Theorems 2.8, 2.11; (2.40), (2.21), and (2.31).   ∎

We present

Theorem 2.36. *All as in Theorems 2.8, 2.11. Assume $\|\omega\|_{L_\infty(\mathbb{R}^n)} < \infty$; $p, q > 1$: $\frac{1}{p} + \frac{1}{q} = 1$, $l > \frac{n}{p}$, $\overline{D}^\alpha f \in L_p(B)$ for $|\alpha| = l$. Then for every $x \in B$ (almost every $x \in B$, respectively), it holds*

$$E(f)(x) \leq \left(\sum_{|\alpha|=l} \left\|\overline{D}^\alpha f\right\|_{L_p(B)} \right) \|\omega\|_{L_\infty(\mathbb{R}^n)} n c_p d^{l-\frac{n}{p}+n} =: A_4. \quad (2.62)$$

Additionally, assume that $\overline{D}^\alpha f \in L_p(B)$, for $1 \leq |\alpha| \leq l - 1$. Then

$$\Delta(f)(x) \leq \left[c_{q,l,n} \left(\sum_{1 \leq |\alpha| \leq l-1} \left(\frac{\left\|\overline{D}^\alpha f\right\|_{L_p(B)}}{\alpha!} d^{|\alpha|+\frac{n}{q}} \right) \right) \right.$$

$$\left. + \left(\sum_{|\alpha|=l} \left\|\overline{D}^\alpha f\right\|_{L_p(B)} \right) n c_p d^{l-\frac{n}{p}+n} \right] \|\omega\|_{L_\infty(\mathbb{R}^n)}$$

$$=: B_4. \qquad\qquad\qquad\qquad\qquad\qquad\qquad\qquad\qquad (2.63)$$

Proof. By Theorems 2.8, 2.11; (2.40), (2.24), and (2.33). ∎

Proposition 2.37. *All as in Theorem 2.35. It holds (for every $x \in B$ and almost every $x \in B$, respectively)*

$$\Delta(f)(x) \leq \|\omega\|_{L_1(B)} \left(\sum_{1 \leq |\alpha| \leq l-1} \frac{d^{|\alpha|} \left\| \overline{D}^\alpha f \right\|_{L_\infty(B)}}{\alpha!} \right)$$

$$+ \frac{n d^{l+n} \pi^{\frac{n}{2}} \|\omega\|_{L_\infty(\mathbb{R}^n)}}{2^n \Gamma\left(\frac{n}{2} + 1\right)} \left(\sum_{|\alpha|=l} \left\| \overline{D}^\alpha f \right\|_{L_\infty(B)} \right)$$

$$=: B_5. \tag{2.64}$$

Proof. By Theorem 2.35 and (2.48). ∎

We also have

Theorem 2.38. *Here all as in Corollary 2.12. Assume $l \geq n$. Then for every $x \in B$ (almost every $x \in B$, respectively), it holds*

$$\Delta(f)(x) := \left| f(x) - \int_B f(y) \omega(y) \, dy \right|$$

$$\leq \left(\sum_{1 \leq |\alpha| \leq l-1} \frac{1}{\alpha!} \left\| D_y^\alpha \left[(x-y)^\alpha \omega(y) \right] \right\|_\infty \right) \|f\|_{L_1(B)}$$

$$+ \left(\sum_{|\alpha|=l} \left\| \overline{D}^\alpha f \right\|_{L_1(B)} \right) \|\omega\|_{L_\infty(\mathbb{R}^n)} \, n d^l. \tag{2.65}$$

Proof. Based on Corollary 2.12, Theorem 2.34 and (2.36). ∎

Theorem 2.39. *Here all as in Corollary 2.12. Suppose $l \geq n$ and $\left\| \overline{D}^\alpha f \right\|_{L_\infty(B)} < \infty$ for all $\alpha : |\alpha| = l$; $f \in L_\infty(B)$. Then for every $x \in B$ (almost every $x \in B$, respectively), we find*

$$\Delta(f)(x) \leq \left(\sum_{1 \leq |\alpha| \leq l-1} \frac{1}{\alpha!} \left\| D_y^\alpha \left[(x-y)^\alpha \omega(y) \right] \right\|_{L_1(B)} \right) \|f\|_{L_\infty(B)}$$

$$+ \frac{n d^{l+n} \pi^{\frac{n}{2}} \|\omega\|_{L_\infty(\mathbb{R}^n)}}{2^n \Gamma\left(\frac{n}{2} + 1\right)} \left(\sum_{|\alpha|=l} \left\| \overline{D}^\alpha f \right\|_{L_\infty(B)} \right). \tag{2.66}$$

Proof. Based on Corollary 2.12, Theorem 2.35 and (2.36). ∎

Theorem 2.40. *Here all as in Corollary 2.12. Assume* $p, q > 1 : \frac{1}{p} + \frac{1}{q} = 1$, $l > \frac{n}{p}$, $\overline{D}^{\alpha} f \in L_p(B)$ *for* $|\alpha| = l$ *and* $f \in L_p(B)$. *Then for every* $x \in B$ *(almost every* $x \in B$, *respectively), we derive*

$$
\Delta(f)(x) \le \left(\sum_{1 \le |\alpha| \le l-1} \frac{1}{\alpha!} \left\| D_y^{\alpha} \left[(x-y)^{\alpha} \, \omega(y) \right] \right\|_{L_q(B)} \right) \|f\|_{L_p(B)}
$$

$$
+ n c_p d^{l - \frac{n}{p} + n} \|\omega\|_{L_\infty(\mathbb{R}^n)} \left(\sum_{|\alpha|=l} \left\| \overline{D}^{\alpha} f \right\|_{L_p(B)} \right). \tag{2.67}
$$

Proof. Based on Corollary 2.12, Theorem 2.36 and (2.36). ∎

We also give

Theorem 2.41. *Here all as in Corollary 2.13 with* $Q_\beta^{l-1} f(x)$ *as in (2.37). Let* $l \ge n + |\beta|$. *Then for every* $x \in B$ *(almost every* $x \in B$, *respectively) it holds*

$$
E_\beta(f)(x) := \left| \left(\overline{D}^{\beta} f \right)(x) - (-1)^{|\beta|} \int_B \left(D^{\beta} \omega \right)(y) \, f(y) \, \mathrm{d}y - Q_\beta^{l-1} f(x) \right|
$$

$$
\le \left(\sum_{|\alpha|=l, \alpha \ge \beta} \left\| \overline{D}^{\alpha} f \right\|_{L_1(B)} \right) \|\omega\|_\infty \, n d^{l-|\beta|} =: A_5. \tag{2.68}
$$

Also, we derive

$$
\Delta_\beta(f)(x) := \left| \left(\overline{D}^{\beta} f \right)(x) - (-1)^{|\beta|} \int_B \left(D^{\beta} \omega \right)(y) \, f(y) \, \mathrm{d}y \right|
$$

$$
\le \left(\sum_{1 \le |\alpha| \le l-|\beta|-1} \frac{1}{\alpha!} \left\| D_y^{\alpha+\beta} \left[(x-y)^{\alpha} \, \omega(y) \right] \right\|_\infty \right) \|f\|_{L_1(B)}
$$

$$
+ \left(\sum_{|\alpha|=l, \alpha \ge \beta} \left\| \overline{D}^{\alpha} f \right\|_{L_1(B)} \right) \|\omega\|_\infty \, n d^{l-|\beta|}. \tag{2.69}
$$

Proof. By Corollary 2.13, (2.25), (2.39), and (2.42). ∎

We continue with

Theorem 2.42. *Here all as in Corollary 2.13. Assume* $l \ge n + |\beta|$; $\left\| \overline{D}^{\alpha} f \right\|_{L_\infty(B)} < \infty$, *all* $\alpha : |\alpha| = l, \alpha \ge \beta$. *Then for every* $x \in B$ *(almost every* $x \in B$, *respectively) we find*

$$
E_\beta(f)(x) \le \left(\sum_{|\alpha|=l, \alpha \ge \beta} \left\| \overline{D}^{\alpha} f \right\|_{L_\infty(B)} \right) \|\omega\|_\infty \frac{n \pi^{\frac{n}{2}}}{2^n \Gamma\left(\frac{n}{2} + 1 \right)} d^{l-|\beta|+n} =: A_6.
$$

$$
\tag{2.70}
$$

Additionally, assume $f \in L_\infty(B)$. *It holds*

$$
\Delta_\beta(f)(x) \leq \left(\sum_{1 \leq |\alpha| \leq l - |\beta| - 1} \frac{1}{\alpha!} \left\| D_y^{\alpha+\beta} \left[(x-y)^\alpha \, \omega(y) \right] \right\|_{L_1(B)} \right) \| f \|_{L_\infty(B)}
$$

$$
+ \left(\sum_{|\alpha| = l, \alpha \geq \beta} \left\| \overline{D}^\alpha f \right\|_{L_\infty(B)} \right) \frac{\| \omega \|_\infty \, n \pi^{\frac{n}{2}}}{2^n \, \Gamma \left(\frac{n}{2} + 1 \right)} d^{l - |\beta| + n}. \quad (2.71)
$$

Proof. By Corollary 2.13, (2.26), (2.39), and (2.42). ■

We finish Ostrowski-type inequalities with

Theorem 2.43. *Here all as in Corollary 2.13. Assume* $p, q > 1 : \frac{1}{p} + \frac{1}{q} = 1$, $\overline{D}^\alpha f \in L_p(B)$ *all* $\alpha : |\alpha| = l, \alpha \geq \beta, l > \frac{n}{p} + |\beta|$. *Then for every* $x \in B$ *(almost every* $x \in B$, *respectively), we derive*

$$
E_\beta(f)(x) \leq \left(\sum_{|\alpha| = l, \alpha \geq \beta} \left\| \overline{D}^\alpha f \right\|_{L_p(B)} \right) \| \omega \|_\infty \, n c_p d^{l - |\beta| - \frac{n}{p} + n} =: A_7. \quad (2.72)
$$

Additionally assume $f \in L_p(B)$. *It holds*

$$
\Delta_\beta(f)(x) \leq \left(\sum_{1 \leq |\alpha| \leq l - |\beta| - 1} \frac{1}{\alpha!} \left\| D_y^{\alpha+\beta} \left[(x-y)^\alpha \, \omega(y) \right] \right\|_{L_q(B)} \right) \| f \|_{L_p(B)}
$$

$$
+ \left(\sum_{|\alpha| = l, \alpha \geq \beta} \left\| \overline{D}^\alpha f \right\|_{L_p(B)} \right) \| \omega \|_\infty \, n c_p d^{l - |\beta| - \frac{n}{p} + n}. \quad (2.73)
$$

Proof. By Corollary 2.13, (2.27), (2.39), and (2.42). ■

We make

Remark 2.44. In preparation to present comparison of integral means inequalities we consider the open ball $B_1 = B_1(y_0, r_1) \subseteq B$. We consider also a weight function $\psi \geq 0$ which is Lebesgue integrable on $\mathbb{R}^n$ with $supp \, \psi \subset \overline{B_1} \subset \Omega$, and $\int_{\mathbb{R}^n} \psi(x) \, dx = 1$. Clearly here $\int_{B_1} \psi(x) \, dx = 1$. For example for $x \in B_1$, $\psi(x) := \frac{1}{Vol(B_1)}$, 0 elsewhere, etc.

We will apply the following principle.

In general a constraint of the form $|F(x) - G| \leq \varepsilon$, where F is a function and G, ε real numbers that all make sense, implies that $\left| \int_{\mathbb{R}^n} F(x) \, \psi(x) \, dx - G \right| \leq \varepsilon$.

Next we give a series of comparison of integral means inequalities based on Ostrowski-type inequalities presented in this chapter. We use Remark 2.44.

Theorem 2.45. *All as in Theorem 2.33. Then*

$$M\,(f) := \left| \int_{B_1} f\,(x)\,\psi\,(x)\,\mathrm{d}x - \int_B f\,(x)\,\omega\,(x)\,\mathrm{d}x \right.$$

$$\left. - \int_{B_1} Q^{l-1} f\,(x)\,\psi\,(x)\,\mathrm{d}x \right| \leq A_1, \tag{2.74}$$

and

$$m\,(f) := \left| \int_{B_1} f\,(x)\,\psi\,(x)\,\mathrm{d}x - \int_B f\,(x)\,\omega\,(x)\,\mathrm{d}x \right| \leq B_1. \tag{2.75}$$

Theorem 2.46. *All as in Theorem 2.34. Then*

$$M\,(f) \leq A_2, \tag{2.76}$$

and

$$m\,(f) \leq B_2. \tag{2.77}$$

Theorem 2.47. *All as in Theorem 2.35. Then*

$$M\,(f) \leq A_3, \tag{2.78}$$

and

$$m\,(f) \leq B_3. \tag{2.79}$$

Theorem 2.48. *All as in Theorem 2.36. Then*

$$M\,(f) \leq A_4, \tag{2.80}$$

$$m\,(f) \leq B_4. \tag{2.81}$$

Theorem 2.49. *All as in Proposition 2.37. Then*

$$m\,(f) \leq B_5. \tag{2.82}$$

Theorem 2.50. *All as in Theorem 2.41. Then*

$$M^\beta\,(f) := \left| \int_{B_1} \psi\,(x)\left(\overline{D}^\beta f\right)(x)\,\mathrm{d}x - (-1)^{|\beta|} \int_B \left(D^\beta \omega\right)(x)\,f\,(x)\,\mathrm{d}x \right.$$

$$\left. - \int_{B_1} \psi\,(x)\,Q_\beta^{l-1} f\,(x)\,\mathrm{d}x \right| \leq A_5. \tag{2.83}$$

Theorem 2.51. *All as in Theorem 2.42. Then*

$$M^\beta (f) \le A_6. \tag{2.84}$$

We finish with

Theorem 2.52. *All as in Theorem 2.43. Then*

$$M^\beta (f) \le A_7. \tag{2.85}$$

2.4 Applications

Example 2.53 (see also [5], p. 93). Let $B := \{x \in \mathbb{R}^n : |x - x_0| < \rho\}$, and

$$\varphi (x) := \begin{cases} e^{-\left(1-\left(\frac{|x-x_0|}{\rho}\right)^2\right)^{-1}}, & \text{if } |x - x_0| < \rho, \\ 0, & \text{if } |x - x_0| \ge \rho. \end{cases} \tag{2.86}$$

Call $c := \int_{\mathbb{R}^n} \varphi (x) \, dx > 0$, then $\Phi (x) := \frac{1}{c} \varphi (x) \in C_0^\infty (\mathbb{R}^n)$ with supp$\Phi = \overline{B}$ and $\int_{\mathbb{R}^n} \Phi (x) \, dx = 1$ and $\max |\Phi| \le const \tan t \cdot \rho^{-n}$.

We call Φ a cut-off function.

One for this chapter's results by choosing $\omega (x) = \Phi (x)$ or $\omega (x) = \frac{1}{Vol(B)}$, etc., can give lots of applications.

Here, selectively we give some special cases inequalities. We start with Chebyshev–Grüss-type inequalities.

Corollary 2.54 (to Theorem 2.26). *All assumptions as in Theorem 2.26. Case of* $l = 1$. *Then*

$$\left| \int_B \omega (x) f (x) g (x) \, dx - \left(\int_B \omega (x) f (x) \, dx \right) \left(\int_B \omega (x) g (x) \, dx \right) \right|$$

$$\le \frac{nd \, \|\omega\|_{L_1(\mathbb{R}^n)}^2}{2} \left[\|g\|_{\infty,B} \|D^\alpha f\|_{\infty,1,B}^{\max} + \|f\|_{\infty,B} \|D^\alpha g\|_{\infty,1,B}^{\max} \right]. \tag{2.87}$$

If $f = g$, then

$$\left| \int_B \omega (x) f^2 (x) \, dx - \left(\int_B \omega (x) f (x) \, dx \right)^2 \right|$$

$$\le nd \, \|\omega\|_{L_1(\mathbb{R}^n)}^2 \|f\|_{\infty,B} \|D^\alpha f\|_{\infty,1,B}^{\max} . \tag{2.88}$$

Corollary 2.55 (to Theorem 2.27). *All assumptions as in Theorem 2.27. Case of* $f = g$, $l = n$ *and* $\omega(x) := \frac{1}{Vol(B)}$, *for all* $x \in \overline{B}$, $\omega(x) := 0$ *on* $\mathbb{R}^n - \overline{B}$. *Then*

$$\left| \int_B f^2(x)\,dx - \frac{2^n \Gamma\left(\frac{n}{2}+1\right)}{d^n \pi^{\frac{n}{2}}} \left(\int_B f(x)\,dx \right)^2 \right|$$

$$\leq \frac{2^n \Gamma\left(\frac{n}{2}+1\right)}{\pi^{\frac{n}{2}}} \left[\left\{ \|f\|_{L_1(B)} \left(\sum_{1 \leq |\alpha| \leq n-1} \left(\frac{d^{|\alpha|-n}}{\alpha!} \left\| D_w^\alpha f \right\|_{L_1(B)} \right) \right) \right\} \right.$$

$$\left. + n \left(\|f\|_{L_1(B)} \left(\sum_{|\alpha|=n} \left\| D_w^\alpha f \right\|_{L_1(B)} \right) \right) \right]. \tag{2.89}$$

We continue the Ostrowski-type inequality.

Corollary 2.56 (to Theorem 2.33). *All as in Theorem 2.33. Case of* $l = 1$. *Then for every* $x \in B$ *it holds*

$$\left| f(x) - \int_B f(y)\,\omega(y)\,dy \right| \leq nd \,\|\omega\|_{L_1(\mathbb{R}^n)} \|D^\alpha f\|_{\infty,1,B}^{\max} := Z_1. \tag{2.90}$$

We finish chapter with a comparison of means inequality.

Corollary 2.57 (to Corollary 2.56). *All as in Corollary 2.56 and Remark 2.44. Then*

$$\left| \int_{B_1} f(x)\,\psi(x)\,dx - \int_B f(y)\,\omega(y)\,dy \right| \leq Z_1. \tag{2.91}$$

References

1. G.A. Anastassiou, *Quantitative Approximations*, Chapman & Hall/CRC, Boca Raton, New York, 2000.
2. G.A. Anastassiou, *Probabilistic Inequalities*, World Scientific, Singapore, New Jersey, 2010.
3. G.A. Anastassiou, *Advanced Inequalities*, World Scientific, Singapore, New Jersey, 2010.
4. G.A. Anastassiou, *Multivariate Inequalities based on Sobolev Representations*, Applicable Analysis, accepted 2011.
5. S. Brenner and L.R. Scott, *The mathematical theory of finite element methods*, Springer, N. York, 2008.
6. V. Burenkov, *Sobolev spaces and domains*, B.G. Teubner, Stuttgart, Leipzig, 1998.
7. P.L. Chebyshev, *Sur les expressions approximatives des integrales definies par les autres prises entre les mêmes limites*, Proc. Math. Soc. Charkov, 2(1882), 93-98.
8. G. Grüss, *Über das Maximum des absoluten Betrages von* $\left[\left(\frac{1}{b-a}\right) \int_a^b f(x)g(x)\,dx - \left(\frac{1}{(b-a)^2}\right) \int_a^b f(x)\,dx \int_a^b g(x)\,dx \right]$, Math. Z. 39 (1935), pp. 215-226.
9. A. Ostrowski, *Uber die Absolutabweichung einer differentiabaren Funcktion von ihrem Integralmittelwert*, Comment. Math. Helv. 10(1938), 226-227.